MW00443440

Estimating
for
Contractors: *How to Make Estimates that Win Jobs*

Paul J. Cook

RSMeans
CMDGROUP

Estimating for Contractors: *How to Make Estimates that Win Jobs*

Paul J. Cook

Illustrations by Carl W. Linde

RSMeans
CMDGROUP

Copyright 1982

R.S. Means Company, Inc.

Construction Publishers & Consultants
Construction Plaza
63 Smiths Lane
Kingston, MA 02364-0800
(781) 585-7880

R.S. Means Company, Inc. ("R.S. Means"), its authors, editors and engineers, apply diligence and judgment in locating and using reliable sources for the information published. **However, R.S. Means makes no express or implied warranty or guarantee in connection with the content of the information contained herein, including the accuracy, correctness, value, sufficiency, or completeness of the data, methods and other information contained herein. R.S. Means makes no express or implied warranty of merchantability or fitness for a particular purpose.** R.S. Means shall have no liability to any customer or third party for any loss, expense, or damage, including consequential, incidental, special or punitive damages, including lost profits or lost revenue, caused directly or indirectly by any error or omission, or arising out of, or in connection with, the information contained herein.

No part of this publication may be reproduced, stored in a retrieval system, or transmitted in any form or by any means without prior written permission of R.S. Means Company, Inc.

The book and cover were designed by Norman R. Forgit. Illustrations by Carl W. Linde. Cover photograph by Norman R. Forgit.

Printed in the United States of America

25   24   23   22   21   20

Library of Congress Catalog Number 82-208250

ISBN 0-87629-271-6

Table of Contents

Part 1
General
Considerations

Part 2
Working
Techniques

Part 2 (cont.)

Part 3
Concluding
Considerations

Part 4
Estimating a
Typical Project

Part 1 General Considerations

1.1 Introduction

The purpose of this book is to educate the reader in all aspects of construction estimating as it is performed within the typical *medium size* construction contracting company (one that grosses 5 to 20 million dollars in contracts yearly). The emphasis will be on *estimating*, with only a minimum regard to quantity surveying on the one hand, and bidding on the other hand. To do justice to those two subjects requires an entire volume for each.

A further aim of this book is to show how estimating may be done with maximum efficiency without at the same time sacrificing quality. Pressure upon the estimator for maximum efficiency comes from the insatiable appetite of the construction company and its never-ending hunger for contracts. With barely enough time to complete his work-on-hand, the estimator is regularly pressed to squeeze another project into his schedule.

This book also proposes to show how the estimator "tailors" his costs to fit his particular company. He does not merely compute the cost to construct a project, he computes what the cost would be for *his particular company* to construct it. He incorporates in his figures, as far as possible, the known production capacities of known people. Beyond that, he works with records of average productions.

All unit costs, tables and charts in this book are only examples and are not intended to be used in the serious estimating of real projects. For current unit prices applicable to real projects, the reader is referred to R.S. Means' *Building Construction Cost Data*. Abbreviations, contractions and terms common in the construction trades are freely used in this book, and for brevity dollar marks ($) and decimal places (0.00), unless they are needed for clarity, are omitted.

1.2 The estimator and estimating

Everyone deals with money and budgets, therefore everyone estimates, but not everyone is an estimator. In a construction company the contractor, the superintendent and others frequently do estimating, but do not think of themselves as estimators. *A person is an estimator when estimating is his main occupation.*

Someone said the word for estimator is "experience". Years of estimating contribute to, but do not guarantee the highest level of experience. Many things contribute: the companies one works for, the minds of the people he works with, the types of construction projects he estimates, the personal responsibilities he carries. As an estimator becomes more experienced he relies increasingly upon his own judgment, which is sometimes facetiously called "educated guessing". It would be dangerous for an inexperienced person to guess, but there may be moments when (as with golfing or driving a car) conscious awareness of technique can, and should be, dropped in favor of "instinct". Often an experienced estimator, after having systematically calculated a cost item, will "know" it is too high or low, and will make a judgmental adjustment (see Sections 1.8—Judgment, 3.5—Contingency Allowances, and 3.6—Presentation Meetings).

Ultimately the proof of expert estimating is partly in the consistency of the bidding, the profit from completed construction projects, and the relative freedom from disputes. It is only partly due to these, because management, superintendence and other factors can also contribute, causing estimates to appear better or worse than they really are.

Estimating is a department within the organization and, depending upon the size of the company, may consist of one person or several (see Section 1.3—The Estimator's Role in the Company). In the medium size company there may be a chief estimator, an estimator or two, and one or more clerical assistants. At peak periods of work, temporary help may be drawn from other persons within or outside the company. The estimating department may be compared to the heart and circulatory system of an organism. It is the dynamic center of a construction company, and its outreach makes it also the heart of the entire construction industry.

Prior to bidding, a new project is of little interest to anyone in a construction company other than the estimator. In order to do his work, the estimator must decide for himself the probable methods of construction procedure. From time to time he may call upon the superintendent or others for advice, but primarily he is on his own regarding such matters as interpretation of specs, choice of equipment, sources of materials, methods of scaffolding, shoring, hoisting, assumptions of concrete form design, conveying and placing of concrete, allowances for waste, methods of dewatering etc. In theory, he constructs the entire project in his own way, in his mind and on paper (see Section 1.14—Selecting Methods of Construction).

Eventually the estimator presents his assumptions, along with cost estimates, to others for criticism and acceptance. Since his ideas are derived from solid experience, most of them survive and some of them undergo revision. More will be said later about the presentation meeting, as it affects bidding and field construction (see Section 3.6—Presentation Meeting). The point is that the estimator's thought processes are influential throughout the entire course of construction, from the moment drawings arrive in the office, through bidding, to the completion of the project in the field and its acceptance by the owner.

The estimator is perhaps better qualified than any other to visualize suitable methods of construction procedure. Because his work exposes him to a greater variety of construction projects than any other person, he has a broad, theoretical education in construction. For every project a company obtains through bidding, the estimator "constructs mentally" ten or more projects which, although they go by the wayside, have been thoroughly incorporated into his reservoir of knowledge.

Estimating is purely mental work, perfectly logical and scientific in form. Its activity occurs between the idea of the designer and the concrete reality by the builder. It interprets the design in a special language called "cost" which, as one aspect of a construction project, has a kind of reality of its own. Ultimately the exact cost will be made known by the bookkeeper, but knowledge of it cannot wait; an approximation is necessary before actual physical work begins. The estimator "reads" the designer's mind and converts the drawing symbols into measurements and man-hours. His are constant choices, made in a consistent manner, toward an imagined whole.

Estimating work requires a special temperament if the estimator is to last long in his field. The mind that is at home with confinement to desk and detail work seems to be in the minority among construction workers. His unique temperament helps the estimator adapt to a working life that rests on an uncertain foundation, where he is always somewhat in error and never more than approximately correct. He literally "wrestles" with numbers. The fact that estimating is *hard work* is one of the reasons for the existence of the occupation. If estimating were nothing more than simple guessing, the owner-contractor could do his own, but estimating is complicated, time consuming, mental effort. Beyond a certain point, however, hard work tends to become unfruitful; once numbers are wrestled into a general relationship, they adjust better to gentler, more intuitive treatment.

If it is true, as some say, that contracting is the greatest gamble in the business world, then the estimator acts as advisor to the biggest gambler. A less harsh way of stating it is: the estimator performs the basic calculations which are used by the contractor as "calculated risks" (see Section 1.8—Judgment).

1.3 The estimator's role in the company

Figure 1.3a

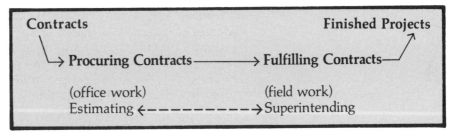

At its simplest a company divides into two departments, as in Figure 1.3a

In a partnership, one partner manages the office and the procurement of contracts, while the other partner is responsible for the field work and fulfillment of the contracts. Information flow is direct between the partners. All construction companies develop around this basic dichotomy. In such a simple partnership, estimating tends to be thrown together for the immediate purpose of bidding; but in a typical medium size company, conditions are provided for estimating to be done as a specialized trade or occupation. Figure 1.3b shows a more complex organization in which, under a manager, one team procures contracts while another fulfills them. Information flow takes place indirectly through the management staff.

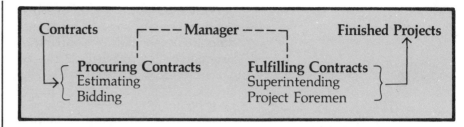

Figure 1.3b

As estimating becomes a departmental appendage to the contract procurement half of the company, a similar effect occurs in the contract fulfillment half, as indicated in Figure 1.3c. The organization may expand in complexity and size, but the two basic responsibilities remain, like nuclei, relatively unchanged.

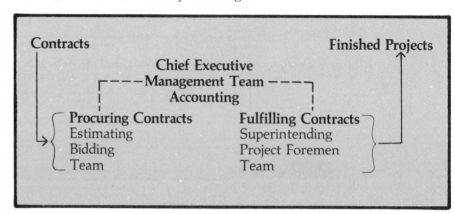

Figure 1.3c

Formally defined, the estimator's role in the company is to compile from drawings, specifications and/or direct observations at construction sites, the data relevent to individual projects; and by certain mathematical conventions, convert the data to proportionate dollar cost values.

It is interesting to note that the employed estimator and superintendent retain a relationship paralleling the original contractor-partners.

1.4
Estimating as a construction cost control

One proof of expert estimating is that it fits within the super's capabilities. As a baseball player is obliged to throw the ball within the catcher's reach, so the estimator's costs need to be within the super's powers to perform.

Once the estimator's work is done (at the moment of bidding) it is entirely up to other people in the company to "prove him right". The super can do a lot to "prove" the estimate, and occasionally to save a poor estimate; but a good way to produce an estimate that is provable and savable is for the estimator to consult with the super when estimating critical cost items. Whether or not he later proves the correctness of the estimate, the super has at least had the chance to influence it.

As construction proceeds an effort is made by the super to accomplish each element within the cost allotted to it by the estimator. The entire project is thus held as closely as possible to and preferably *under*, the estimated cost. Most cost overruns are due to laxity in this system of control. It is not just a monitoring to see how the costs are going; it is positive project management. The super and the estimator continually update their knowledge of their own capabilities and those of the workmen and equipment. Consequently, not only are present projects in process of construction being cost-controlled, future projects will be more accurately and competitively estimated and more dependably cost-controlled (see Section 3.12—Feedback).

1.5 Objectivity in estimating

An estimate approximates some as yet unrealized (future) cost of a project. Because of many variables a project has the potential of costing more or less than some ideally estimated amount. Fig. 1.5a shows the tendency for a project, as it runs over or under the assumed ideal cost, to approach a theoretical high or low point. These are not absolute limits, but the shadings indicate that a loosely or tightly managed project will rise, or be reduced in cost only so far, and as the limits are approached any efforts to raise or lower the costs further become increasingly difficult.

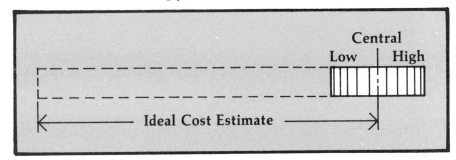

Figure 1.5a

At the low point, specifications, quality standards and production efficiency limitations resist cost cutting; at the high point, the resistance is in the opposition by the owner, the contractor, or both, to loss of capital.

Whereas the actual cost of a project will fall within a plastic high/low range, faulty estimating may produce a figure that is outside the range entirely (too high or too low). The aim of the estimator is to "hit the target" as centrally as possible. When pressures to be either competitive or cautious influence the estimate, the range itself, does not change, but the estimate draws nearer one limitation or the other, as in Figure 1.5b.

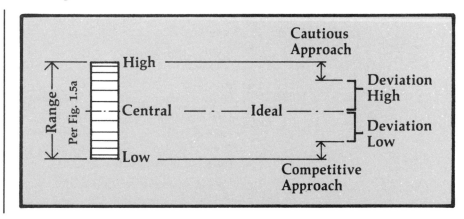

Figure 1.5b

This high/low principle is also present within each item and element, for instance; concrete work, forming, carpentry, doors, etc. It may be presumed that if the estimate for each element successfully hits the central point of its range, the total estimated cost of the project will be central. But while such an achievement is not probable, there is an averaging effect in the estimating process which tends to achieve the same desired result. Figure 1.5c illustrates this averaging effect. A large number of elements, symbolized by A,B,C and D, with estimates hitting randomly above and below centrality, will tend, by cancellation, to a sum which is central (or very near it).

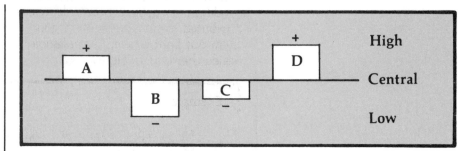

Figure 1.5c

This can occur only if the estimator works with a disinterested (objective) attitude toward the outcome of the bidding. A strong influence toward either cautiousness or competitveness can cause the total estimated amount to fall outside the range of limitations, as shown in Figure 1.5d.

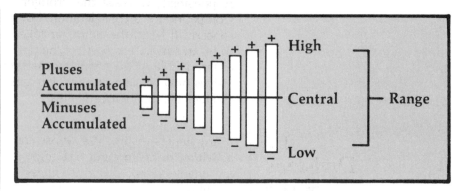

Figure 1.5d

Objectivity, then, is a necessity for accurate estimating; and it is entirely up to the estimator to achieve it.

1.6 Considering competition

Objectivity, thoroughness and accuracy are the estimator's goals. Adjustments for competitiveness may come later, as the moment for bidding approaches, and they may be made with or without his involvement.

A very desirable project (see Section 2.4—Analyzing a Project for Desirability) may cause material quantities to be taken closer, more speculative methods of construction selected, and a higher than average level of labor production assumed. In other words, "the best way", and the project would be figured "tight". A less desirable project might cause slack in the estimate, all calculations based upon the "worst", or most pessimistic way, and the project would be figured "loose" (see Figure 1.5d).

If uninfluenced, the estimator remains objective at all times; but with a sense of realism, he may lean gently in the direction that competition requires of his company. He may legitimately lean to the safer or to the riskier side so long as he keeps clearly in mind the relationship of his estimate to the purely objective.

The estimator may select a factor (such as 5%) and apply it to each unit price, thus raising or lowering his entire estimate by that proportion. Prior to bidding the mood might change for any of several reasons; the trend, due to competition, might influence the estimate upward or downward. Whatever happens, adjustments would be easy to make without losing sight of the purely objective viewpoint.

When influenced to revise his figures, the estimator should keep a record of his objective cost estimates, because memories of the bidding conditions fade, and months later cost underruns may be unjustly attributed to his judgment. A good way to keep records is to cross them out lightly (keep them legible) and print the revised figures beside them, as in Figure 1.6a.

Figure 1.6a

Example:					
			.06		960
Fine grade for slabs	16,000 sf	@	~~.07~~	=	~~1,120~~
Hand excavate	120 cy	@	17.00	=	2,040
			5.50		451
Backfill	82 cy	@	~~7.00~~	=	~~574~~

Paradoxically, the estimator, more than any other person, has the responsibility to resist the stronger influences toward competitiveness. Competition is a consideration, but it is not the estimator's primary concern. If later, the estimator takes charge of the bidding, he ceases to be an estimator; in effect, he puts on a different hat, assumes a different role in the company. As estimator, he is a technician and strives only for accuracy; his job is not to be low bidder, but to produce correct cost estimates.

1.7 Divisions of the estimate

Many general estimators whose companies have strong subcontracting activities, never attempt to budget the costs of specialized trades. They are unable, or unwilling, to "guess" with reasonable accuracy the costs of subtrades. This defines one of the main subdivisions: *the general contractor's own direct on site labor, material and equipment costs.* Because he estimates this portion of every project, the estimator is an expert in the several trades that make up his company's field (earthwork, concrete work, carpentry, etc.). Due to his personal experience and aptitudes he is more proficient in some trades than others. His greatest expertise is at the center. He becomes less sure toward the perimeter of his abilities and increasingly dependent upon subcontractors. See Figure 1.7a.

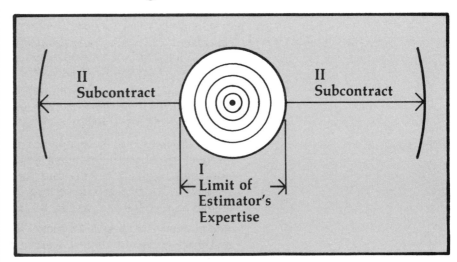

Figure 1.7a

At this perimeter (dotted line) begins the second major classification: *portions of a project normally let to subcontractors.* See Figure 1.7b

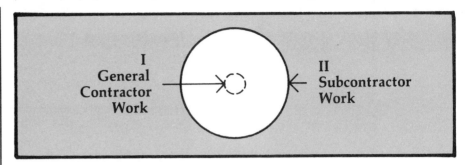

Figure 1.7b

Within his own field the estimator's knowledge is not often challenged; it may be challenged when he presumes to estimate the cost of trades that lie outside his field of specialization. Yet, there are many occasions when he is required to be so presumptuous. The results are not estimates; they are *budgets* (see Section 3.1—Budgeting Subtrades), and at the least, plug-in figures (see Section 3.4—Plug-in Figures and Allowances).

A complete estimate is made by "breaking down" a construction project into lesser and lesser pieces which, for convenience, may be named trades, categories, items and elements. Briefly, a breakdown for estimating looks something like the example in Figure 1.7c. Generally, as the pieces become smaller there are more elements than items, more items than categories, and more categories than trades. The estimator's interest in a sub-bid (main division II) is the total dollar amount, the scope of coverage and the assurance of its guarantee. Most of this information comes not while the estimate is being made, but in the later process of bidding.

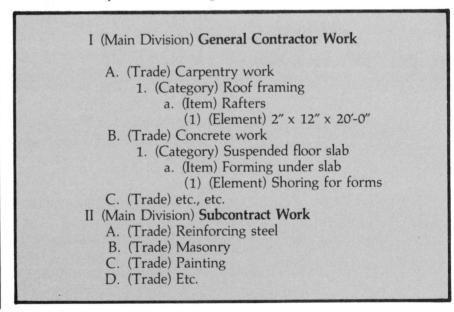

I (Main Division) **General Contractor Work**

  A. (Trade) Carpentry work
    1. (Category) Roof framing
      a. (Item) Rafters
        (1) (Element) 2" x 12" x 20'-0"
  B. (Trade) Concrete work
    1. (Category) Suspended floor slab
      a. (Item) Forming under slab
        (1) (Element) Shoring for forms
  C. (Trade) etc., etc.
II (Main Division) **Subcontract Work**
  A. (Trade) Reinforcing steel
  B. (Trade) Masonry
  C. (Trade) Painting
  D. (Trade) Etc.

Figure 1.7c

The project may be divided into even smaller pieces than customary— estimators vary as to how much breaking down they do. Theoretically, there is a practical limit to the minuteness of a breakdown. The process is usually stopped at the level where it no longer contributes to the accuracy of the estimate.

As a rule the estimator breaks down a project into more detail than a person who is not an estimator. The estimator knows more cost-effecting considerations, both those that add to and those that reduce the cost. Non-estimators tend to "lump" quantities together and overlook important cost factors such as layout, distribution of materials, reuses or salvage value of materials, cleanup, etc.

The relationship to the whole estimate of each work item (its critical value) dictates the degree of breakdown. Obviously, an item of little cost value (such as a single concrete catch basin) would not require as much breakdown as an item of large cost value (such as a concrete water reservoir).

The degree of breakdown is usually controlled by the estimator's sense of satisfaction as he seeks an accurate cost level. In general, the more cost records that are available to him, the less breaking down he needs to do, because breaking down is the means to arrive at unkown lump sums. When lump sums are known, and proved in the field, breaking down is unnecessary.

Figure 1.7c takes the vertical form; but horizontal expansion also takes place, with columns for labor, material, equipment, subcontract and total, as in Figure 1.7d.

Figure 1.7d

	Example of vertical and horizontal expansion				
Description	**Labor**	**Mat'l**	**Equip**	**Subcont**	**Total**
Trade					
Category					
Item					
Element					

Remembering that the object of estimating is finding the cost (dollars), some lumping together is advisable. In other words, to a degree the end justifies the means. Items of small quantity having the same unit price may be combined. *The term "item" is used generally to represent any one of the subdivisions* as in Figure 1.7e.

Figure 1.7e

Example:

Longitudinal construction joints (LCJ)
$$2,000 \text{ lf @ } 1.25 = 2,500$$
Transverse construction joints (TCJ)
$$1,600 \text{ lf @ } 1.25 = 2,000$$
Combine the above
$$\text{LCJ \& TCJ} \quad 3,600 \text{ lf @ } 1.25 = 4,500$$

1.8 Judgment

There is no element in estimating that does not include a touch of judgment (guessing has a slightly different meaning). No unit price is absolute. All cost records from different projects show variations for the same item. This is probably due to the fact that field conditions are rarely identical. Two or more estimators will conceive different costs for the same unit of work. Construction companies will differ in their conceptions of costs. An estimator who changes employment finds that he must change many of his preconceived ideas of cost.

Although it is true that one company may consistently produce specific items of work at a different cost from another company, even within that company individuals differ in their convictions about costs.

One of the duties of the estimator is to resolve the various conceptions of cost. In this respect he is like a judge who considers all the "evidence" placed before him.

His deliberations are usually rapid and he is unaware of them; but occasionally he takes the time to study in depth unusual or critical cost items.

Although judgment is unavoidable, it is usually brought into play following all the factual, firm and scientific methods. The estimator tries to minimize the neccessity for judgment. There are degrees of certainty and dependability in an estimate, listed in order from strongest to weakest as follows:

1. Firm bid from subcontractor, or quotation from material supplier.
2. Material or equipment on hand (owned by contractor and thus not subject to change).
3. Type of construction is common practice and estimator is on familiar ground.
4. Type of construction is occasionally practiced and estimator is not unfamiliar with it.
5. Type of construction is rare and estimator is not familiar with it.
6. Type of construction that estimator knows nothing about.

Every estimate will have one or more of the above conditions. Their number and severity determine how much the estimator's judgment is exercised. A good way to approach them is by isolating the uncertain cost items and boxing them in red pencil to draw attention, so that special thought and research may be provided.

A tug-o'-war always exists between the need to be competitive and the need to be certain. Greater competitiveness is accomplished by reducing certainty and relying more on judgment (see Section 1.6—Considering Competition). In this context, judging and guessing are almost synonymous. For instance, instead of using a firm quotation, the estimator uses a price which he believes material can be bought for (based upon recent purchases). He may be doing a bit of both judging and guessing, and he may also be taking a "calculated risk", knowing that a firm quotation is not necessarily the lowest price obtainable.

At this point a distinction should be made between the terms calculated risk and gamble. Figure 1.8a indicates the difference as a degree of risk.

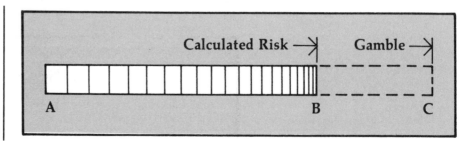

Figure 1.8a

Calculated risk begins at *A* with a mild dependence on chance, where the probability greatly favors the desired outcome. The risk increases in the direction of *B* where the odds are 50-50. From this point on toward *C* chance increasingly weighs against the desired outcome. Ideally an estimate approaches *A* as nearly as possible.

1.9 Sources of cost information

Typical sources of cost information, roughly in order of reliability are:

1. Records of actual costs repeatedly proved in the field.
2. Tables and mathematical relationships.
3. Cost reference books.
4. Constructive imagination.
5. Hearsay information.

Cost record systems are used by the more exacting construction companies. They vary from simple spot-checking to sophisticated, time consuming, costly operations (see Section 3.12—Feedback). Although the best of them are only guidelines, cost records are the most reliable of cost information sources. A good system shows all items in the project and requires the super to account for all of them. A portion of a simplified example is given in Figure 1.9a.

Figure 1.9a

| Item | | Estimated | | | Field Cost Record | |
| | Quantity | Labor | | | Actual | Actual |
		Unit	Cost		Cost	Unit
Hand excavation	210 cy	22.00	4,620		4,788	22.80
Footing forms	3,200 sf	2.40	7,680		7,072	2.21
Footing concrete	135 cy	7.15	965		907	6.72

The main value of the field cost record is the confirmation and further refinement of the estimator's calculations. A comparison of cost records from different projects assures the estimator of reasonable accuracy in pricing. If, for instance, six projects had recorded footing forms (of similar design) per square foot as: 1.97, 2.15, 1.93, 2.13, 2.29, and 1.88, the average of 2.06 would be a convincing unit price to use. However, since it is only an average, it should be adjusted for special conditions in a project presently being estimated (see Section 1.16—Keeping up with Changes).

Tables and mathematical relationships are convenient for adjusting average unit prices. In the following example, Figure 1.9b, of a table of footing forms, 2.06/square foot is shown as average (up to a point, labor unit costs decrease as form materials are reused).

Figure 1.9b

| | Number of Uses | | | | |
	1	2	3	4	5
Cost/sf	2.49	2.27	2.06	1.85	1.67

Adjustments may also be made for quantity. Assume that records show the placing of concrete in footing forms to cost an average of 6.17/cubic yard, use of such a table as in Figure 1.9c could advise an increase or decrease.

Figure 1.9c

	Quantity				
	50cy	75cy	100cy	150cy	200cy
Cost/cy	7.47	6.79	6.17	5.56	5.00

The elevation above or below ground level affects unit costs because of variations in handling, hoisting, climbing, reaching, etc. High concrete formwork costs more than low formwork because more material and labor are involved to strengthen and brace them against the increased pressure.

Cost reference books offer the estimator information on unfamiliar items. Distilled from the experiences of numerous experts, book information is representative of typical cases and average conditions. It is a valuable tool to help with marginal estimating problems, to "sell" the estimate, and to make budgets of subtrades (see Section 3.1— Budgeting Subtrades).

Often second and third opinions are worthwhile, and when two or more reference books are consulted, their values may be compared to aid in the selection of unit prices. Figure 1.9d provides an example in which the selected unit prices are underlined.

Figure 1.9d

	Unit Costs Per Square Foot		
	Reference Books		Estimator's
	A	B	Records
Forming concrete columns	2.60	2.82	2.70
Placing concrete in columns	14.65	19.00	22.50
Forming concrete beams	2.95	3.10	2.75

In this example, neither the highest nor lowest unit price was used. In actual practice, the estimator might prefer his own unit prices exclusively, but with some adjustments as influenced by the reference books. Figure 1.9e is a sample page from R.S. Means' *Building Construction Cost Data*.

Cost reference books benefit the estimator indirectly by preparing the minds of others with whom he must deal, such as architects, engineers, construction executives, inspectors, superintendents, persons in government agencies, etc., who are prepared to accept "ballpark" figures as offered by an independent, third party; thus, much time is saved in hashing over unit costs. However, the stage is thereby set for strenuous debating in cases where the estimator's figures differ considerably from the book values.

In cases of construction items that are lacking in cost information, the estimator employs his imagination to do the work visually, recording

3.3	CAST IN PLACE CONCRETE	CREW	DAILY OUTPUT	UNIT	BARE COSTS			TOTAL INCL O&P
					MAT.	INST.	TOTAL	
102	Average reinforcing	C-17B	11	C.Y.	134	145	279	355
104	Maximum reinforcing		7.40		263	220	483	595
120	Columns, round, spirals, 16" diam., minimum reinforcing		9.90		261	165	426	515
122	(51) Average reinforcing		6.20		371	260	631	775
124	Maximum reinforcing		4.60		491	350	841	1,025
130	20" diameter, minimum reinforcing		13		251	125	376	450
132	Average reinforcing		7.20		362	225	587	715
134	Maximum reinforcing		5.30		467	305	772	945
140	24" diameter, minimum reinforcing		15.80		235	105	340	405
142	Average reinforcing		9.70		322	165	487	590
144	Maximum reinforcing		6.70		455	240	695	840
150	36" diameter, minimum reinforcing		24.20		192	67	259	305
152	Average reinforcing		16.60		242	98	340	405
154	Maximum reinforcing		9		355	180	535	645
170	Curbs, 6" x 18", straight	C-15	185	L.F.	2.65	6.45	9.10	12.15
175	Curb and gutter	"	170	"	4.40	7.05	11.45	14.85
190	Elevated slabs, flat slab, 100 psf L.L., 20 ft. span	C-17A	15	C.Y.	98	100	198	250
195	30 ft. span	C-17B	22		96	74	170	210
210	Flat plate, 100 psf L.L., 15 ft. span	C-17A	14		107	110	217	270
215	25 ft. span	C-17B	24		94	68	162	200
230	Waffle const., 19" domes, 100 psf L.L., 20 ft. span		13.50		123	120	243	305
235	30 ft. span		13.10		118	125	243	305
250	One way joists, 20" pans, 100 psf L.L., 15 ft. span	C-17A	10		146	150	296	375
255	25 ft. span		12		140	125	265	335
270	One way beam & slab, 100 psf L.L., 15 ft. span		10.80		112	140	252	325
275	25 ft. span		10.70		129	140	269	345
290	Two way beam & slab, 100 psf L.L., 15 ft. span		9.70		129	155	284	365
295	25 ft. span	C-17B	14.70		116	110	226	280
310	Elevated slabs including finish, not							
311	including forms or reinforcing							
315	Regular concrete, 4" slab	C-8	1,850	S.F.	.61	.69	1.30	1.58
320	6" slab		1,650		.91	.77	1.68	2.02
325	2-1/2" thick floor fill		2,175		.36	.58	.94	1.17
330	Lightweight, 110# per C.F., 2-1/2" thick floor fill		1,900		.46	.67	1.13	1.39
340	Cellular concrete, 1-5/8" fill, under 5000 S.F.		2,000		.22	.64	.86	1.08
345	Over 10,000 S.F.		2,200		.19	.58	.77	.97
350	Add per floor for 3 to 6 stories high		31,800			.04	.04	.05L
352	For 7 to 20 stories high		21,200			.06	.06	.08L
380	Footings, spread under 1 C.Y.	C-17B	18.50	C.Y.	80	88	168	210
385	Over 5 C.Y.	C-17C	45.70		76	38	114	135
390	Footings, strip 18" x 9", plain	C-17B	22.90		64	71	135	170
395	36" x 12", reinforced		37.60		69	43	112	135
400	Foundation mat, under 10 C.Y.		20.30		71	80	151	190
405	Over 20 C.Y.		27.20		65	60	125	155
420	Grade walls, 8" thick, 8 ft. high	C-17A	9.60		102	160	262	340
425	14 ft. high		7.40		129	205	334	435
426	12" thick, 8 ft. high		13.90		86	110	196	250
427	14 ft. high		11.20		96	135	231	300
430	15" thick, 8 ft. high	C-17B	18		79	90	169	215
435	12 ft. high	C-17A	14.70		82	105	187	235
450	18 ft. high	"	12.20		94	125	219	280
451								
465	Ground slab, not including finish, 4" thick	C-17C	47.40	C.Y.	60	36	96	115
470	6" thick	"	61	"	58	28	86	105
475	Ground slab, incl. troweled finish, not incl. forms							
476	or reinforcing, over 10,000 S.F., 4" thick slab	C-8	2,025	S.F.	.65	.63	1.28	1.54
482	6" thick slab		1,900		.95	.67	1.62	1.93
484	8" thick slab		1,750		1.27	.73	2	2.35
490	12" thick slab		1,350		1.87	.94	2.81	3.30
495	15" thick slab		1,250		2.34	1.02	3.36	3.91

Figure 1.9e

the motions, the number of workmen, the type of equipment, the time periods, and thus, the cost. All construction projects have points in common; the estimator uses those points which he knows in new and different combinations.

Constructive imagination runs throughout the work of estimating. To begin with, imagination is used to break down a project into its lesser subdivisions, for example, a concrete slab on ground. In his mind the estimator "sees" the elements of fine grading, edge forms, aggregate base course, membrane waterproofing, setting of screeds, laying reinforcing mesh, spreading the concrete, troweling and curing. He imagines the number of men of different trades and payscales, the time they will spend completing each element of work, and the equipment they will use.

Hearsay information may be considered from persons who claim to have special experience. The dependability of such information rests upon the confidence the estimator has in the individuals who supply it.

1.10 Forms, formats and systems of estimating

The forms used by estimators vary slightly from company to company, but over many years, estimating forms have become more or less standardized. They quicken the understanding between people, provide easy recall of details of past projects, encourage respect and acceptance of the estimate by persons who recognize good workmanship, help to avoid errors of omission and provide backtracking when necessary of any items in the estimate. Typically the forms used by an estimator are:

1. *Specifications Summary Sheet,* Figure 1.10a. This is a listing of all trades and categories in the project. It serves the purposes of:

   a. Acquainting the estimator with those trades which he will have to figure himself.
   b. Providing a thorough outline to be used later in putting the bid together.
   c. A guide to the soliciting of sub-bids.
   d. Identifying the trades affected by alternate bids.
   e. A means of discovering and recording cost items that might otherwise be overlooked.

The sample form shows the outline by trades of a typical project having a base bid and three alternate bids. In the alternate bids some trades are additive and others deductive (see Section 2.15—Alternate Bids and Bid Schedules).

2. *Blank Pad for Miscellaneous Notes,* Figure 1.10b. As the estimating work proceeds, the estimator discovers ambiguities, discrepancies, omissions, etc., and makes notations of them for future investigations. As he continues to work, some of the questions are answered, and he crosses them off the list. When he has completed the estimating work, any unanswered questions are then taken up with those persons most concerned: A & E, owner's agent, subcontractor, project superintendent, etc. In the sample form a few typical notes are shown to illustrate its use.

3. *Quantity Survey (take-off) Forms,* Figures 1.10c and 1.10d. It is unlikely that one form design will serve equally well to take off all kinds of quantities; earthwork, concrete, carpentry, specialties, etc.; however, the sample form may serve fairly well. Figure 1.10c shows several items and elements of a concrete take off. Figure 1.10d is a special form for the listing of many associated elements, too numerous for the shorter form to accommodate.

4. *Price-out Form,* Figure 1.10e. This form may be used to price-out all trades. The sample contains a few examples of price-outs to show how the form is used. Notice how the lesser subdivisions are indented.

   In this form the material column doubles to include equipment, to save space on the sheet. Many estimators prefer to use a legal size sheet containing a separate column for equipment, which makes possible a clearer analysis of both the material and the equipment.

5. *Worksheet,* Figure 1.10f. Each of the items and elements on the price-out form, Figure 1.10e, is followed by simple unit prices and extensions, and very little explanation. For those which need more explanation, the worksheet, Figure 1.10f, may be used. The resulting calculations are then transferred to the price-out sheet, where they become a matter of record. The worksheet may be the same form as the price-out sheet, but a different color paper would identify it, to avoid confusion. The use of the worksheet is demonstrated in Part 4 — Estimating a Typical Project.

Figures 1.10g, 1.10h, 1.10i, 1.10j and 1.10k are MEANSCO forms providing a choice between three different summary sheets and two quantity sheets.

Specifications Summary Sheet

Project: **Science Lab** Bid Date: **9/18/82**

Location **Long Beach, CA** Calendar Days to Complete **600**

Spec. Section	Trade	Base Bid	Alt. #1	Alt. #2	Alt. #3
1A	General Conditions	X*	—	—	—
2A	Demolition	X*	Add	—	—
2B	Earthwork	X*	Add	—	—
2C	Asphalt conc. paving	X	Add	—	—
2D	Chain link fence	X	Add	—	—
2E	Irrigation & landscaping	—	X	—	—
3A	Concrete work	X*	Add	—	—
3A	Reinforcing steel	X	Add	—	—
4A	Masonry	X	Add	—	—
5A	Structural & misc. metal	X	Add	—	—
6A	Carpentry & millwork	X*	—	—	—
7A	Roofing & roof insulation	X	—	—	—
7B	Wall & ceiling insulation	X	—	—	—
7C	Sheetmetal work	X	Add	—	—
7D	Caulking & sealants	X*	—	—	—
8A	Finish hardware	X	—	—	—
8B	Metal doors & frames	X	—	—	—
8C	Metal windows & glazing	X	—	—	—
8D	Steel overhead doors	X	—	—	—
9A	Painting & wall covering	X	Add	—	—
9B	Ceramic tile	—	—	Add	—
9C	Lathing & Plastering	X	—	—	—
9D	Metal framing & gyp. bd.	X	—	—	—
9F	Resilient flooring	X	—	Deduct	—
9G	Carpeting	—	—	Add	—
9H	Acoustical tile	X	—	—	—
9K	Glazed wall coating	X	—	Deduct	—
10A	Toilet partitions	X	—	—	—
10B	Bldg. specialties	X	—	—	—
10E	Portable partitions	X	—	—	—
11A	Adjustable loading ramp	X	—	—	—
11B	Dental casework & equip.	X	—	—	—
12A	Fixed theater seats	X	—	—	—
13A	Radiation protection	X	—	—	—
13B	Walk-in refrigerator	X	—	—	—
13C	Integrated ceiling	X	—	—	—
14A	Hydraulic elevator	X	—	—	—
15A	Mechanical work	X	Add	—	Add
16A	Electrical work	X	—	Deduct	—

Figure 1.10a

* indicates the trades which will be figured by the general estimator. Spaces marked with "X" will receive the total basic amounts; others are merely additive or deductive amounts, and spaces not to receive anything are lined out.

Notes and Questions Sheet

Project: ____Science Lab____ Bid Date: __9/18/82__

	Note	Proposed Action	Resolution
1.	Sec. 1A.04 fire insurance is required	obtain fee	ok
2.	See Sec. 1A.17 regarding 25% of work to be done by prime contractor	noted	ok
3.	Required time of completion 600 calendar days-may not be long enough	make a progress schedule	use 645 days
4.	Will dewatering be required below elevation = 5'?	investigate	yes
5.	Where is nearest disposal site for debris?	investigate	5 miles
6.	Will soil stand vertical, or what is angle of repose?	study soil reports	will stand
7.	Details C/C-7 & 20/A-18 are in conflict.	check with architect	use C/C-7
8.	What kind of finish to be used on balcony slab?	check with architect	non-slip
9.	Use Spanall-type forms under suspended slabs?	check with super	yes

Figure 1.10b

Quantity Survey Sheet

Project: **Science Lab**   Bid Date: **9/18/82**

Detail Footings		L	W	D	Conc.	Form	Exc.	Disp.
		Dimensions			**Extensions**			
1/s4	perim.							
	conc.	630	1.34	1.0	844	630	—	—
	exc.	630	1.34	2.0	—	—	1,688	—
	disp.	630	1.34	1.5	—	—	—	1,266
2/s4	conc.	240	2.0	1.0	480	240	—	—
	exc.	240	2.0	2.0	—	—	960	—
	disp.	240	2.0	1.67	—	—	—	802
5/s7	inter.							
	conc.	118	1.34	1.0	158	118	—	—
	exc.	118	1.34	1.25	—	—	198	—
	disp.	118	1.34	1.0	—	—	—	158
	Totals				1,482 sf	988 lf	2,846 cf	2,226
	change to cy (÷27)				55 cy		106 cy	83 cy

Floor Slab on Ground					SF	Conc.	4" Gravel	Membr.
4"	thick	42.34	172	.34	7,282	2,476	2,476	7,282
6"	thick	80.0	172	.5	13,760	6,880	4,678	13,760
	Totals				21,042 sf	9,356 cf	7,154 cf	21,042 sf
	change to cy (÷27)					347 cy	265 cy	

Walls Above Grade				Conc.	Form	Finish
7/s4	12.0	1.0	15.0	180	360	360
	4.0	1.0	7.0	28	56	56
	6.0	1.0	5.0	30	60	60
	15.0	1.0	15.0	225	450	450
	6.0	1.0	5.0	30	60	60
	Totals			493 cf	986 sf	986 sf
	change to cy (÷27)			19 cy		

Figure 1.10c

Note: These and all other quantities are gathered together and transferred to the price-out sheet.

Quantity Survey Sheet

Project: _____   Bid Date: _____

Detail	Dimensions									
	L	W	D							

Figure 1.10d

Price-out Sheet

Project: _Science Lab, Long Beach, CA_   Bid date: __9/18/82__

Detail / Description	Production	Quantity	Units L	Units M	Labor	Material	Subcontr.	Total
Concrete Cast-In-Place								
Footings - layout	18 mhrs	988 lf	.29	.05	287	49	---	336
mach. excav.	8 hrs @ 55.00	106 cy	---	4.15	---	440*	---	440
hand excav.	42 hrs	1,482 sf	.37	---	548	---	---	548
forming	164 mhrs	988 lf	2.49	.55	2,460	543	---	3,003
concrete	32 mhrs	55 cy	8.15	50.00	448	2,750	---	3,198
backfill	10 mhrs	23 cy	6.09	3.04	140	70*	---	210
disposal	4 hrs @ 50.00	83 cy	---	2.40	---	199*	---	199
Slab on grd. - fine grade	96 mhrs	2,142 sf	.06	---	1,263	---	---	1,263
gravel base	92 mhrs	265 cy	4.86	12.00	1,288	3,180*	---	4,468
concrete	140 mhrs	347 cy	5.65	50.00	1,961	17,350	---	19,311
set screeds	86 mhrs	21,042 sf	.065	.015	1,368	316	---	1,684
trowel/cure	344 mhrs	21,042 sf	.26	.026	5,471	547	---	6,018
Walls - forming	170 mhrs	986 sf	2.59	.60	2,554	592	---	3,146
concrete	18 mhrs	19 cy	13.26	60.00	252	1,140*	---	1,392
rub & grind	46 mhrs	986 sf	.65	.08	641	79	---	720

*These material items include operated equipment

Figure 1.10e

Work Sheet

Project: _____ Bid Date: _____

Detail	Description	Production	Quantity	Units		Labor	Material	Subcontr.	Total
				L	M				

Note: See Part 4 for example of usage of this form.

Figure 1.10f

1.11 Interpretations and disputes

Plans and specs, if perfect, would permit only *one* clear and indisputable interpretation. In reality they fall short of that ideal. Most professionally drawn plans present only occasional problems of interpretation and many of those, if cost is unimportant, are easily resolved; difficult to resolve problems occur when the cost is considerable.

The estimator is not so much concerned with avoiding disputes as having solid foundation for his own interpretations. The burden rests on him to decide if he has enough information, and this points to his first principle: *the estimator has a right to assume a literal interpretation.* He is not required to second guess the designer's intentions. If his interpretation makes sense to him, he must proceed with it. If it does not make sense to him, he should bring the question to someone's attention (see Figure 1.10b) and keep a record of the investigation.

Most problems of interpretation offer more than one choice. Rather than request clarification on many slightly vague questions, the estimator relies upon his experience with the attitudes of architects, inspectors and others in the construction industry (see Section 1.21— Customs-of-the-Trades). Designers and draftsmen leave much to the intellects of the interpreters. The estimator assumes that elements of a project of great importance to the designer will be appropriately detailed and described. Generally, omission of detail is a tacit permission for the estimator to use his best judgment.

Interpretations of drawings and specs involve assumptions; it is up to the estimator to minimize them and become expert in distinguishing between "safe" and "unsafe". Customs-of-the-Trades provide guidelines to help interpret the drawings and specs. Some of them are:

1. Specifications take priority over drawings.
2. Large scale details take priorty over small scale.
3. Dimensions take priority over scaling.

The estimator learns to observe closely the meanings of words, phrases and punctuations. The location, presence or absence of a comma can make a great difference. In disputes, the owner's agents on the one hand, and the contractor's on the other, tend to defend the interpretations that are advantageous to themselves. The estimator is often caught in the middle. If negotiations fail, a disinterested authority (court, or board of arbitration) decides the case; and questions such as these are asked: should the estimator have interpreted differently? Should he have studied the issue longer, deeper? Should he have investigated, asked questions?

When a dispute involves an oversight, the question arises: How thoroughly should the specs have been read by the estimator? There are four levels of reading, roughly in this order, from the most thorough downward:

1. Sections concerning the trades which the prime contractor will do directly with his own forces, relying on the estimator's calculations.
2. Sections of general and special conditions which, although mostly standard, may contain critical instructions.
3. Sections concerning trades which the prime contractor might not be able to sublet, and the estimator will roughly estimate (budget).
4. Sections concerning trades which subcontractors will be responsible for.

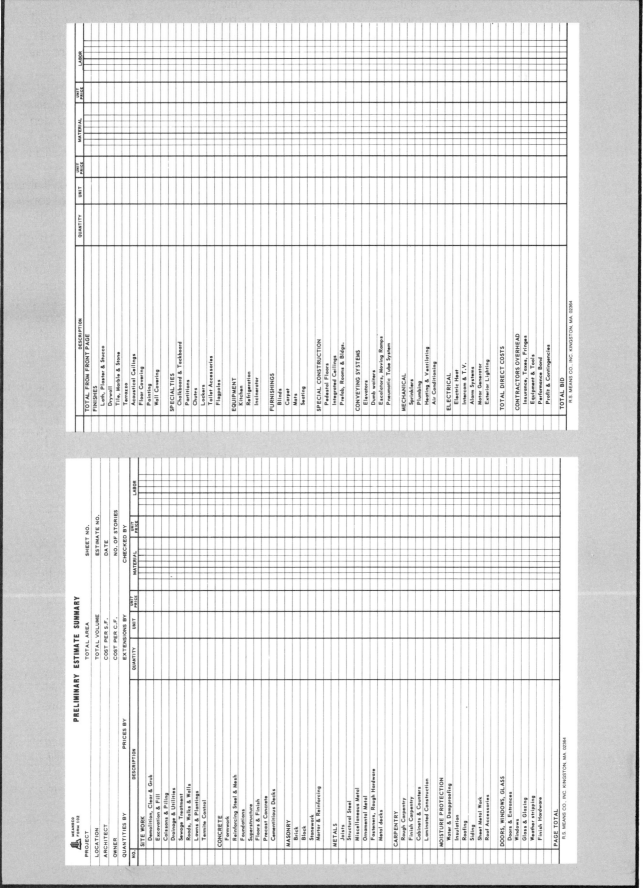

Figure 1.10g

MEANSCO FORM 115

CONDENSED ESTIMATE SUMMARY

PROJECT	TOTAL AREA	SHEET NO.
LOCATION	TOTAL VOLUME	ESTIMATE NO.
ARCHITECT	COST PER S.F.	DATE
OWNER	COST PER C.F.	NO. OF STORIES
QUANTITIES BY	PRICES BY	EXTENSIONS BY	CHECKED BY

NO.	DESCRIPTION	MATERIAL	LABOR	SUBCONTRACT	TOTAL	ADJUSTMENT
	SITE WORK					
	Excavation					
	CONCRETE					
	MASONRY					
	METALS					
	CARPENTRY					
	MOISTURE PROTECTION					
	DOORS, WINDOWS, GLASS					
	FINISHES					
	SPECIALTIES					
	EQUIPMENT					
	FURNISHINGS					
	SPECIAL CONSTRUCTION					
	CONVEYING SYSTEMS					
	MECHANICAL					
	Plumbing					
	Heating, Ventilating, Air Conditioning					
	ELECTRICAL					
	TOTAL DIRECT COSTS					
	CONTRACTORS OVERHEAD					
	Performance Bond					
	Profit & Contingencies					
	TOTAL BID					

R.S. MEANS CO., INC. KINGSTON, MA. 02364

Figure 1.10h

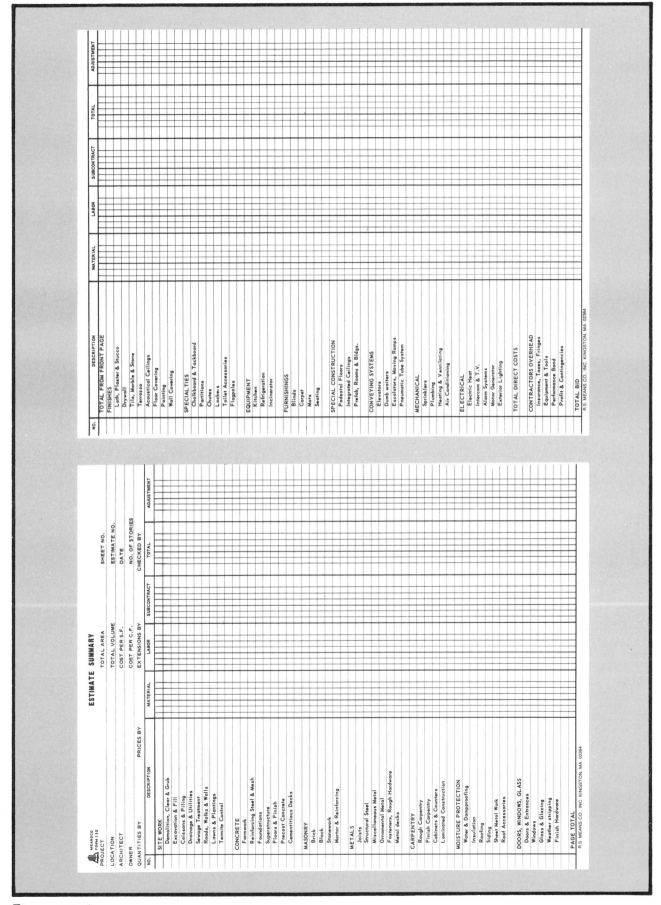

Figure 1.10i

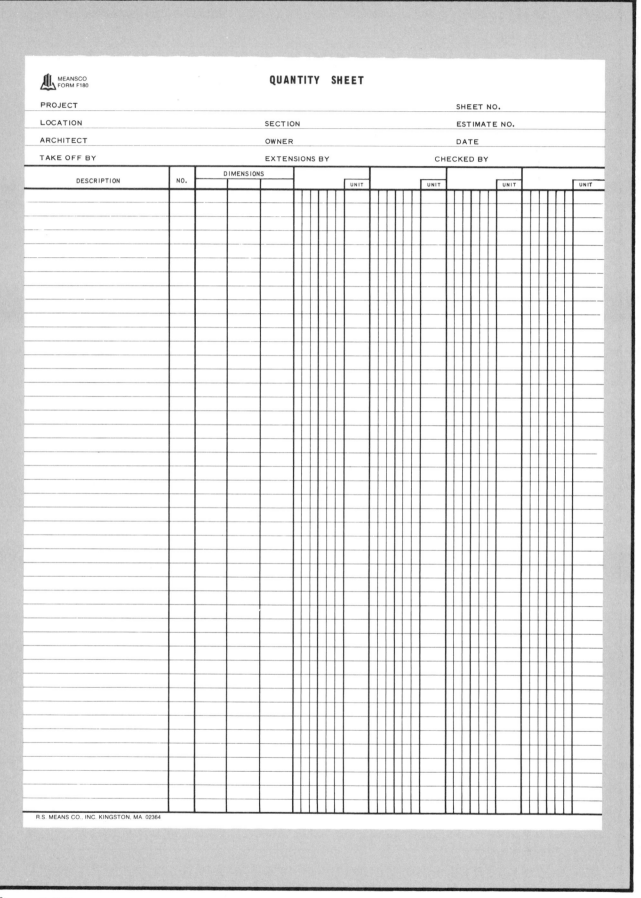

Figure 1.10j

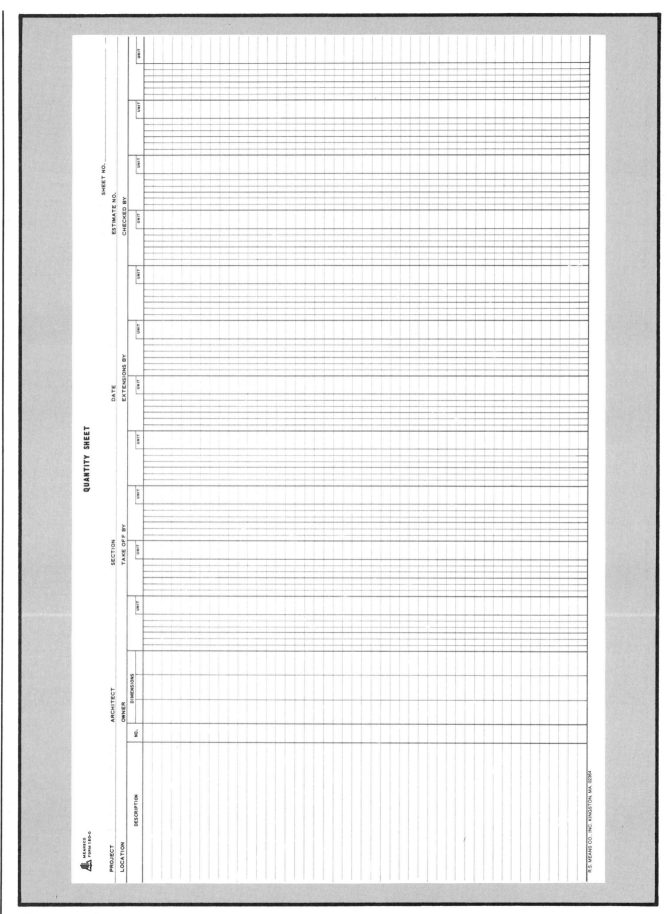

Figure 1.10k

The fourth and least thoroughly read level is important only for ascertaining the "coverage", and points of separation between trades. The estimator makes the first, cursory, survey of the specs for coverage, in order to know the scope of his own necessary estimating work. Knowledge of those subcontract sections of the specs accumulates as the project nears the bid deadline, and discussions occur with prospective sub-bidders. Such knowledge pertains more to the business of bidding than estimating.

To summarize, the estimator is required to read the *entire* specifications, but with varying shades of thoroughness, as necessary to competently do his own, personal work. If in his opinion the drawings and specs are inferior, requiring more than a normal amount of guess work, the estimator should inform others in the company who might consider the risk of bidding.

1.12 Ethical considerations

In his objective role, the estimator is not much troubled with problems of ethics. Ethics become a serious problem later in the bidding phase. In the estimating phase, ethics is confined mostly to honoring the confidentiality of material quotations. The estimator is expected to obtain several competitive quotations on such materials as pre-mixed concrete, lumber, plywood and rough hardware; keep the bidder's prices secret from one another; and use whichever prices he considers best (not necessarily the lowest); see Section 3.3—Dealing with Material Suppliers. Early conversations with competing subcontractors are expected to be on an equal-information basis (see Section 3.2—Dealing with Subcontractors).

It is only when he serves the double role of estimator and bidder, or when influenced (see Section 1.6—Considering Competition) that the estimator is faced with ethical choices, simply because bidding adds the dimension of company need for contracts. Bidders seek *legitimate* advantages over their competitors. But not all advantages are legitimate; some are illegal and others are unethical, and the classification is not always quickly apparent. For that reason, every possible advantage is examined. The owner sets the stage for this when he invites bids; he desires and expects bidders to be competitive.

Bidders have too much expense in time, energy and money to perform casually; and yet it is up to each bidder to draw his own line on both the recognition of and obedience to ethical principles.

The contracting business is more vulnerable than other businesses to problems of ethics, perhaps due to the secrecy in which a great volume of quotations are tendered. Opportunities abound for the bending, stretching and abusing of rules of fair play. However, the estimator (as independent of the bidder) does not ordinarily strive to produce the lowest possible cost estimate; and consequently, he may select only the clearly legitimate quotations. It should be candidly acknowledged here that the estimator, although he may be scrupulously ethical, works in an opportunistic social environment. Caught between givers and seekers of information, he may be an occasional victim of circumstances. It is as important for him to master this aspect of his occupation as that of measuring and computing from drawings.

The estimator is responsible (as is a person in any other occupation or profession) only for his own judgment and workmanship. As a salaried person, he has nothing personal to gain or lose from the estimating/bidding process. This fact, coupled with a need for an objective attitude, tends to help him avoid ethical problems.

There is cause, however, for the estimator to be a highly responsible person: his work involves more judgment than most jobs, and forms the basis for his company's financial commitments. He is an important factor in the success or failure of his company. This is a moral as well as professional responsibility, and conscientious fulfillment of it is the estimator's main ethical consideration.

1.13 Mistakes, errors and slips

Webster's Seventh New Collegiate Dictionary defines mistakes as: "A wrong action or statement proceeding from faulty judgment, inadequate knowledge or inattention. *Error* may imply carelessness or willfulness in failing to follow a true course or a model, but it may suggest an inaccuracy where accuracy is impossible; *mistake* implies misconception or inadvertence and is seldom a harsh term; *blunder* commonly implies stupidity or ignorance and usually culpability; slip carries a strong implication of inadvertence or accident producing trivial mistakes; *lapse* implies forgetfulness or inattention."

Blunders and lapses are hardly applicable to the work of an experienced estimator, but mistakes, errors and slips crop up in every estimate, just as they do in championship archery, or other target sports. They are the estimator's occupational hazard. He strives always to prevent them, at best succeeds in only minimizing and living with them, yes, and learning from them. Seldom does he make exactly the same kind of mistake twice. Numerous though they may be, mistakes, errors and slips may be grouped under four headings:

1. Mistakes of judgment
2. Mistakes of omission
3. Mistakes of arithmetic
4. Slips

If every detail of a project were correctly understood (judged), included (not omitted), correctly computed, and free from slips, the estimate would be faultless. An estimator feels that he has achieved that perfection with an occasional project, but if the contract is not obtained, or feedback not supplied, he never learns the truth.

Generally, *mistakes of judgment* are the most difficult to justify or correct. They are slips of the mind, wrong assumptions. A mistake of interpretation (judgment), when the drawings and specs are not misleading, is the estimator's fault, and as a rule he is not forgiven them. The individual is held accountable for his own reasoning and logic. Unit prices are matters of judgment and are binding in contracts; they are regarded as the estimator's sincere intentions, and guaranteed by the bidder. Wrong choices of construction methods and procedures are mistakes of judgment, and the bidder is responsible. They are usually not discovered until the actual construction work is nearly finished, because they are experimental, and can only be credited to the contractor's "education". Bidders are permitted only foresight, and are not given (unless the drawings and specs are faulty) the later advantages of hindsight.

Mistakes of omission occur when an element, item, category, or trade is inadvertently overlooked and left out of the estimate. Entire structures, portions of structures, or all the work on a sheet of drawings, section of specifications or addendum may be overlooked. The causes of such mistakes are seemingly infinite in variety; therefore, there is no simple strategy for their prevention. In Part 4 of this volume, which takes the reader through the estimating procedure of a hypothetical construction project, some methods to minimize all classes of mistakes are shown.

Mistakes of omission are the most to be criticized. It is better to have the value of an item misjudged, or have errors in its arithmetic, than have nothing at all allowed for it.

Mistakes in arithmetic are the easiest to prevent. There are ways that the estimator may check his calculations as he goes along, but because it is human nature to repeat the same mistake, it is a good policy to have all figures checked by another person.

When mistakes of omission and mistakes in arithmetic have been minimized by every possible strategy, the estimator's mind is free to concentrate on the remaining source of error-judgment, to isolate the main problems and cope with them.

Mistakes which cannot be easily classified under the above three types are *slips*. They include such mistakes as the following:

Transpositions
Misquoting
Mishearing
Mismeasuring
Misnoting
Misreading
Forgetting

A few guidelines for preventing mistakes are:

1. Use good techniques for deciding the advisability of bidding a project in the first place. The biggest mistake of all is bidding the wrong job (see Section 2.13—Analyzing a Project for Desirability).
2. Use a good system of notations rather than depend upon memory. Place items that might be forgotten in conspicuous places on the paper where they cannot be overlooked.
3. Check drawing sheets and spec pages to see that none are missing.
4. Check for duplicate buildings, structures, floor plans, etc., represented by a typical drawing or detail.
5. Use a system of marking off and shading (filling in with colored pencil) the areas on the drawings as they are taken off. Take off a quantity first, before shading it in. Anything not shaded will show up plainly, and thus omissions will be avoided.
6. Make a space in the price-out sheet for every item and element requiring a cost figure, and put something, if only a dash, in every space; then look for blank spaces to avoid omissions.
7. Shade with colored pencil detail designations where the quantities are taken off, thus: $\left(\frac{5}{12}\right)$;

   Mark through designations that are seen but not taken off, because they pertain to subcontractor work, thus: $\left(\frac{11}{12}\right)$.

8. Whenever possible, avoid carrying figures forward from sheet to sheet. Confine groups to one page in order to prevent errors in carrying forward subtotals.
9. Avoid deductive costs, or show them in red pencil.
10. When using a calculator or adding machine, anticipate the extensions, using rough mental arithmetic, to catch errors. In other words, learn to know if an arithmetical product "looks" correct.

11. Read the specifications repeatedly—at least three times; the first time cursorily, to get the general concept; second, researching as the estimating proceeds; finally, an overall reading for perspective.
12. In general, learn to see detail without forcing judgments. Let images take shape naturally, but loosely. Avoid arguing a viewpoint, but remain open to a change of mind.
13. Look particularly for mistakes on the short side. It is easier to underestimate than overestimate; and it is more dangerous.

After all possible has been done to avoid mistakes, those that nevertheless occur are seldom fatal to the profit margin of a contract. When the estimating work is completed there remain a few days for discovering errors and correcting them before the bid deadline and its final committment. No mistake is really made until the moment the bid is tendered; until then, everything in the estimate is tentative, and any mistakes that may exist have a possibility of being discovered.

1.14 Selecting methods of construction

It is possible for an entire estimate to be made without consideration of particular methods or equipment. Unit prices may be used exclusively without regard to special conditions; but that approach is not recommended. Yet, an estimator may easily waste time in the consideration of methods of construction. Unlike the super who must do the actual construction work, the estimator's choices are only tentative; it is enough for him to visualize reasonably efficient, economical and feasible methods upon which to base the estimate. Whatever the first assumptions of methods may be, they are subject to later revisions in light of discoveries, reflections and consultations.

To get on with his work, the estimator does well to select customary, proved methods, and avoid the more imaginative. Unless the savings are significant, an unconventional method of construction may not be worth the risk. Some portions of projects are unique, so that familiar methods are not applicable; in such cases the estimator must virtually originate methods which satisfy his sense of logic.

Here are a few examples of questions and choices that confront the estimator:

Demolition-where to dispose of debris; will there be a dumping fee; will burning be permitted; salvage allowance; shoring of floors, roof and walls which are to remain; noise and dust control; type of equipment to use; safety provisions.

Earthwork-type of equipment to use; whether blasting is permitted; source of borrow material; whether existing material meets the specifications for fill; where to dispose of excess material; drainage control; need for dewatering; dust control; where to stockpile material for later use on site; access to working areas; order of procedure; amount of over-excavation for structures; whether shoring is required; responsibility for surveying and engineering.

Concrete work-what kind of forming, metal or wood, gang forms, flying forms, slip forms; number of reuses of form material; working room for prefabrication of forms; method of conveying concrete to forms, crane, pump, belt, buggy, direct pour from trucks; testing and inspection requirements; extremes of temperature provisions; types of finishes; need for scaffolding, and what kind; safety provisions.

Precast concrete work-will building floor slab serve for casting the units; number of different shapes, as they will affect stacking and reuses of forms; sizes and weights for choosing hoisting equipment, is there working room around and overhead; method of bracing shoring and securing.

Carpentry and millwork-storage and protection of material; prefabrication and precutting; hoisting; shoring and scaffolding; conveying materials by elevators, etc; assembly line methods.

Subcontract work-order of procedure, scheduling; shoring, scaffolding and hoisting; storage and working room; coordination; quality control.

Estimating the cost of construction is more accurately done if the estimator, being specific in the selection of equipment, "constructs" his unit costs rather than "derives". A simple example is:

Derived (from cost records)
excavate, load on trucks and haul off-site
1200 cy @ 1.80 = 2,160.00

Constructed (by selected equipment)

1 cy backhoe	1 ea @ 75.00	=	75.00/hr
10 cy dump trucks	3 ea @ 55.00	=	165.00/hr
	Total		240.00/hr

production: 4 cycles per truck per hour = 120 cy/hr

240/120 = 2.00/cy, and 1200 cy @ 2.00 = 2,400.00

To summarize: in general and up to a point, the more an estimator imaginatively selects methods and equipment with which to "construct" his unit prices, rather than use the units "derived" from cost records or other sources, the more dependable the estimate.

1.15
Mobilization and time lag allowance in labor

A labor cost includes *all* the time associated with an item of work. For example, assume that a wall cabinet is to be installed; the time (labor) includes such elements as: studying the drawings for location in the room, the dimensional layout, method of fastening, moving the cabinet from storage, securing it to walls and floors, installing doors, drawers, hardware, top and final trimming. The number of cabinets, their lengths, heights, complexity, etc., affect the unit cost. A large number of similar cabinets develop crew efficiency, and long cabinets require fewer motions per linear foot than shorter cabinets. For example: a single base cabinet, twenty four feet long, might cost 108.00, or 4.50 per linear foot, while eight similar cabinets, three feet long each might cost 144.00, or 6.00 per linear foot.

Non-estimators tend to make no allowance for preparation, lost motion, correction of mistakes, rest breaks, hunting for tools, hardware, helpers, etc. They think of the *net* time to do an element of work, when all materials are on hand and the workmen know exactly what to do. But estimators must think in terms of *gross* time, and the difference between net time and gross time is loosely called "mobilization" time. The greater the number of repetitive elements of work (to a point) the less mobilization time is involved. An illustration may help make this principle clear; in Figure 1.15a the ■'s represent elements of work accomplished, and the spaces between them represent mobilization time. The total cost is the same for each case (A,B,C), but the unit costs differ.

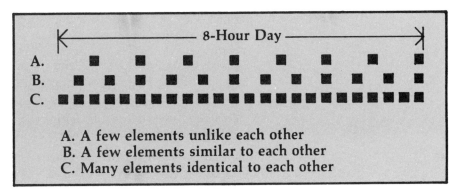

Figure 1.15a

A. A few elements unlike each other
B. A few elements similar to each other
C. Many elements identical to each other

It is more important for the estimator to determine the amount of work that men *will probably do* in a day than what they are theoretically capable of doing.

In projects requiring many months to complete, the estimator must consider the effect of pay raises (see Section 1.16—Keeping up with Changes). He tries to avoid a time lag by figuring the probable production at the (future) time that the work will actually occur. For example, if a certain type of concrete formwork costs 2.15 per square foot currently, and pay scales will be 6% higher at the time that the work will be performed, the unit price will increase to: 2.15 + 6% = 2.28 per square foot; and if roof framing will be done when labor will be 8-1/2% higher than today's 325.00 per thousand board feet, the unit price will be: 325.00 + 8-1/2% = 353.00 per thousand board feet.

What has been said above applies to normal conditions; unusual conditions occur when portions of a project will be under construction in extremes of temperature, rains, winds or tides. Easily overlooked is the standby time, when men attempt to work, but accomplish little.

Weather's effect on almost any outdoor project is usually minor, and is allowed for in the unit prices. Climate is a different subject (winter, summer, desert, mountain, ocean, rainy season, etc.); its effect on production may be so great that unit costs become ridiculous. It is better to use normal units costs throughout the estimate and add a separate cost item for "hardship" conditions; for example:

Assume that the normal cost of placing and vibrating concrete is 5.20 per cubic yard, but because of climatic conditions, workmen are paid per day (eight hours) while producing only the equivalent of five hours of work, the unit price would be:

5.20 x 8/5 = 8.32 per cubic yard

Instead of doing the great volume of work required to adjust every unit price, the following method is preferred:

Estimate the cost of a project using normal unit prices and adjust the labor by categories. Climatic conditions may not affect a project consistently throughout its span of time, and a progress chart may appear as shown in Figure 1.15b on the following page.

In this example, climate affects all of the earthwork, all of the formwork, one half of the concrete work, none of the carpentry or other work, at the rate of five productive hours in eight. The estimated increase in labor cost over normal production for all of the work considered was 37.3%. In addition to simplifying the estimating work, this method has the advantage of providing a comparison of the total difference in cost between the normal and abnormal productions.

	Normal Production	Adjustment			Revised Labor
Earthwork	6,500		6,500 × 8/5	=	10,400
Formwork	140,000		140,000 × 8/5	=	224,000
Concrete (½)	67,000	$\dfrac{67,000}{2}$ +	$\dfrac{67,000}{2}$ × 8/5	=	87,100
Carpentry	24,000				24,000
Other work	52,000				52,000
Totals	289,500				397,500

Figure 1.15b

1.16 Keeping up with changes (inflation)

Rapid and continuing inflation makes obsolete the cost records of projects as recent as a year old; furthermore, the unit costs applicable to the starting months of a project are obsolete for the ending months. Because costs of different items increase at different rates, an estimate incorporates several different methods of compensating for inflation, some of which are listed as follows:

1. Material suppliers are requested to guarantee their quotations for the duration of the contract, or for specified time periods.
2. Materials are purchased before price increases and stored until needed, provided storage space is available and not too expensive.
3. Consideration is given to the substituting of materials, alternate to those specified, if the specified ones are expected to increase in price.
4. Labor increases are negotiated with trade unions at specific dates and rates, so that they may be computed rather than approximated.
5. More consideration than usual is given to the subletting of trades and thus reducing risk.

The increased risk due to inflation is distributed down the line of suppliers and subcontractors.

Estimates are composed of *future costs* and, inflation causes increased "guesstimating". To explain in detail how the estimator adjusts for inflation, the three columns on the price-out sheet are taken one at a time as follows:

1. *Labor.* For example, assume an 18-month project—the easy way is to use an average increase over the entire span, as in Figure 1.16a.

   By trade/categories, carpenters doing concrete formwork in the early months of the project may receive, say, 16.21 per hour, causing only a 5% increase in the formwork unit costs, and receive 16.98 in the late months of the project, causing a 10% increase in rough carpentry unit costs, as in Figure 1.16d.

Figure 1.16a

Projected wages	(18 months in the future)	20.85/hr
Current wages	(average of the trades)	−18.13/hr
	Difference	2.72/hr

Increase spread over 18 months — = $\dfrac{2.72}{2}$ = 1.36/hr

Average rate to use throughout project = 18.13 + 1.36 = 19.49/hr

Percentage increase to apply to all unit costs = $\dfrac{19.49}{18.13}$ =1.075 = 7½%

A more accurate way is to use the actual dates and periods of pay increases, as in Figure 1.16b.

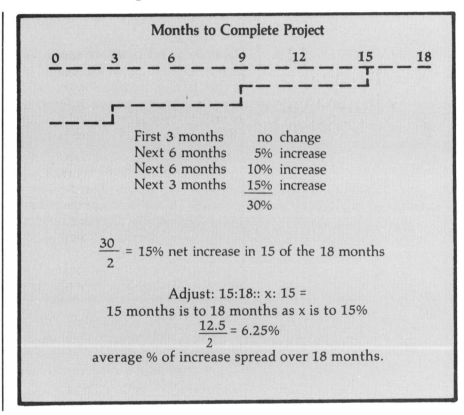

Months to Complete Project

First 3 months	no change
Next 6 months	5% increase
Next 6 months	10% increase
Next 3 months	15% increase
	30%

$\dfrac{30}{2}$ = 15% net increase in 15 of the 18 months

Adjust: 15:18:: x: 15 =
15 months is to 18 months as x is to 15%
$\dfrac{12.5}{2}$ = 6.25%
average % of increase spread over 18 months.

Figure 1.16b

To be more accurate still, adjust for the periods in which actual work will be performed, as in Figure 1.16c.

Unit costs occurring in	1st 3 months - no increase	1.60/sf
	next 6 months - increase 5%	1.68/sf
	next 6 months - increase 10%	1.76/sf
	next 3 months - increase 15%	1.84/sf

Figure 1.16c

Figure 1.16d

	Current Unit Cost	Increase	Use
Forming cost	2.40/sf	5%	2.52/sf
Framing cost	320.00/sf	10%	352.00/sf

Figure 1.16d

2. *Material*. Rather than rely on recent price quotations, the estimator requests new quotations for each project. In some cases when supply is limited, strikes pending, or other causes preclude firm quotations, the estimator may request from suppliers their "educated guesses".

Materials, such as pre-mixed concrete, composed of several ingredients of which one is soon to be increased in price, may be calculated as in Figure 1.16e.

Quotation:

6 sk mix, price good for 3 months:		51.20/cy
after 3 months add (per sk of cement)		1.25/sk
after 12 months add another		1.40/sk

The estimator has the choice of:

(1) **Applying an average increase;**

$$51.20 + \frac{6\,(1.25 + 1.40)}{2} = 59.15/cy$$

(2) **Applying the entire increase;**

$$51.20 + \quad 6\,(1.25 + 1.40) \quad = 67.10/cy$$

(3) **Applying the increases incrementally;**

1st 3 months	150 cy @ 51.20 =	7,680
Next 12 months	1,800 cy @ 58.70 =	105,660
Next 3 months	300 cy @ 67.10 =	20,130
Totals	2,250 cy @ 59.32 =	133,470

Figure 1.16e

Note in this example how nearly the average method and the incremental agree, indicating that the easier method of averaging is to be preferred. The incremental method would be better in projects where a large percentage of the concrete would be placed in either the early or in the late months.

3. *Equipment.* Continuing inflation tends to make more desirable the ownership, rather than rental of equipment. But whether owned or rented, equipment costs are usually computed by the hour, including operators (labor) and fuel. Consequently, increases in rental rates involve two elements: the bare machine rental rate, and the operating engineer (labor) rate. A regular inquiry of rental rates is advisable; then, a projection of probable increases may be approximated, as in Figure 1.16f. For simplicity, let the rental rate include fuel as well as maintenance and repair.

Assume current rental rate of operated 45 ton hydro-crane is 120.00/hr, of which operator and rigger/oiler is 54.00. Expected labor increase is 7% and equipment increase (bare rental) is 4%, then:

Crane	(120.00 - 54.00) + 4% =	68.64/hr
Labor	54.00 + 7% =	57.78/hr
Use for estimating		126.42/hr

Figure 1.16f

1.17 Workmanship in estimating

In estimating, as in other occupations, favorable appearance of the work is an asset. Because of time limitations, constant revisions and expense, typing of the estimating work sheets is not practical. As they are hand printed by the estimator, so they are seen, preserved, stored away and referred to for several years.

Legibility is the most important goal. Estimating may involve some guesswork, but there is no excuse for anyone to have to guess at the identity of numbers and letters set down on the paper by the estimator.

Organization is of next importance, the use of headings (preferably in capital letters), underlinings, indentures, orderliness and numbered sheets.

Consistency is valuable for swift and easy communication and for reference. The same forms and formats (see Section 1.10—Forms, Format and Systems of Estimating), used in the same systematic manner for all projects, compiles a library of well thought out and tested cost data.

The effort to do neat workmanship also tends to produce accuracy and thoroughness. Methodicalness in the hand slows the mind and provides time for second thoughts.

Nice appearing work stimulates in the observer an attitude of confidence toward the estimator and his product. It is assumed that good workmanship connotes pride, and that the basis for pride is quality.

Although a reasonably high standard of workmanship should apply to every estimate, conditions may justify some variations; for instance; informal, routine and private work may not warrant the same care for appearance as public and formal presentations.

1.18 Estimator's working conditions (environment)

The essence of estimating work is concentration. Rude breaks in concentration are opportunities for errors; chosen breaks (taken at "stopping points") are opportunities for second thoughts, and for checking, correcting, revising and reorganizing. Because of inflexible bid deadlines, estimating follows a schedule (see Section 2.14— Scheduling the Time to Estimate). For this reason, privacy is necessary to the estimator.

The room in which the estimator works should be restricted for another reason: to protect the drawings and work sheets from disarrangement or removal. The room should be large enough for several tables, to display more than one set of drawings at a time. It should be well lighted and ventilated. A good arrangement, Figure 1.18a, is one that provides privacy for the estimating team and, under

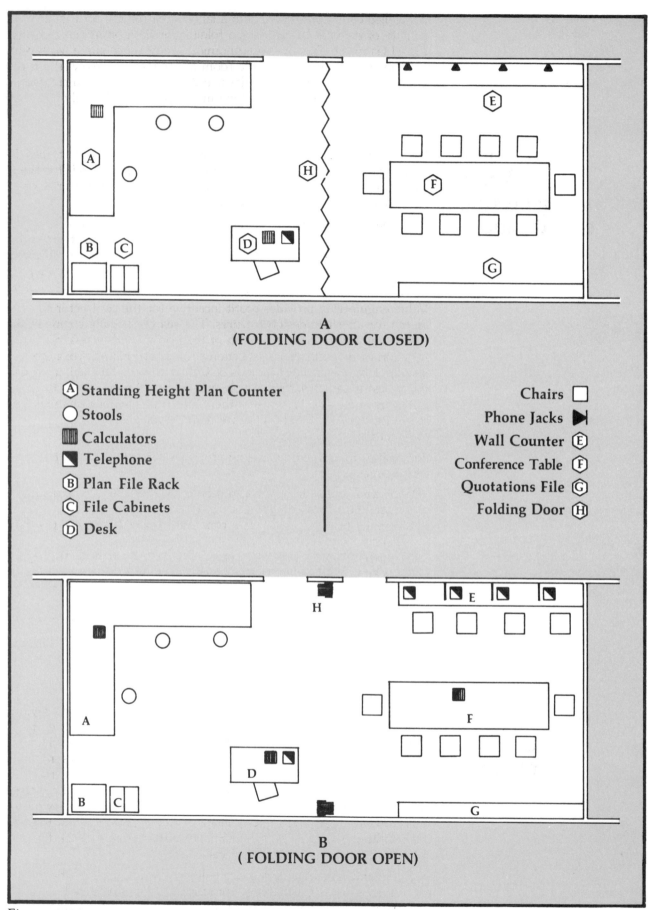

A
(FOLDING DOOR CLOSED)

Ⓐ Standing Height Plan Counter

◯ Stools

▦ Calculators

◤ Telephone

Ⓑ Plan File Rack

Ⓒ File Cabinets

Ⓓ Desk

Chairs ☐

Phone Jacks ▶

Wall Counter Ⓔ

Conference Table Ⓕ

Quotations File Ⓖ

Folding Door Ⓗ

B
(FOLDING DOOR OPEN)

Figure 1.18a

controlled conditions, opens into a larger area for the occasional activity of bidding. In this plan, a folding partition ordinarily remains closed (as in A). This makes economical use of office space, with one portion serving as a conference room for general company use. It can be expanded (as in B) for bid days, and for any period of increased activity requiring additional estimating personnel; only slight rearrangement is necessary.

1.19 Value engineering considerations

For the purpose of fair competition, all bidders are required to bid the drawings and specs as they are. However, it is a good habit for the estimator to search for alternative, more economical, functional, or time-saving features to propose later.

Presumably, the designer is aware of the various methods of construction and specifies the best way. Presumably, also, taste and personal preference may support the designer's choice, causing him to reject other design advantages. But innovations, or sudden changes in comparative prices might bring about worthwhile new considerations.

Value engineering provides profit incentive for the contractor to search for alternate design features. The search logically begins in the estimating phase; as an extension of the normal and necessary selection of methods of construction basic to estimating work (see Section 1.14—Selecting Methods of Construction). The estimator uses those best methods which comply with the requirements of the drawings and specs. However, other (value engineering) methods requiring the A & E's approval are retained for post-bidding consultations.

Possibilities for value engineering proposals may be found in the following examples:

1. *Site work* — change grades to balance cuts and fills, thus saving cost of imports and exports.
2. *Wall construction* — substitute concrete block for concrete; precast concrete for concrete block.
3. *Suspended floors* — substitute precast concrete, pan joist, or steel decking with light weight concrete fill.
4. *Concrete work* — Substitute precast for cast-in-place; or cast-in-place for precast.
5. *Subcontract work* — Encourage subcontract specialists to suggest value engineering ideas; particularly the major trades, such as structural steel, mechanical and electrical.

1.20 Adjusting for complexity/ simplicity

The more complex the project the more likelihood for under-estimating such things as time to complete, general conditions, layout, supervision, etc. The simpler the project the greater the chance of over-estimating. The reason for this is the estimator's perspective from a base of average complexity. "Average thinking" biases all estimating; in effect it tries to make all projects average. As a rule, competitive bids on a project of average complexity are usually closely grouped, while those on either a complex or simple project may vary widely in range. Knowing this principle, some estimators tend to over-compensate, producing overly optimistic sums on simple projects; pessimistic sums on complex projects.

A systematic estimating procedure will incorporate most of the complexity/simplicity variations from average. Yet, an overall (judgment) factor may be in order. Experience shows that simple projects tend to be cut-and-dried, offering little opportunity for "pickups", while complex projects are just the opposite. Allowing for this aspect is speculative, and is usually reserved for the last consideration in bidding (see Section 3.5—Contingency Allowances).

1.21 Customs-of-the-trades

By tradition there are many procedures understood and practiced by the various tradesmen in the construction industry. They are called "customs" and have a more or less enforceable authority in themselves. They are, to say the least, guidelines for determining the responsibility for items of work and standards of workmanship.

The estimator relies upon his knowledge of the customs-of-the trades to include costs to install materials which are usually furnished by subcontractors, such as bolts, inserts, miscellaneous metal, specialty items, etc. He knows where to draw the line between items of work customarily performed by one trade or another, and by the general contractor's employees. The guidelines, as well as his knowledge of the strengths and weaknesses of subcontract activities in his locality, tell him which items to budget for cost, (those which are not clearly specified under the work of particular subcontractors), or which customs-of-the-trades have not categorized.

Following are a few typical borderline items, not always clearly delegated by customs-of-the-trades to particular subcontractors:

Trades	Borderline Items
Demolition	Moving of buildings Structural demo (remodeling) Coring or saw cutting of pavement
Earthmoving	Structural exc (pits, backfill) Shoring and dewatering Aggregate subbase Weed or termite treatment
Landscaping	Topsoil Concrete mow strips
Asphalt concrete	Parking striping & bumpers Header forms Traffic control
Reinforcing steel	Dowels; welding Hoisting; scaffolding
Masonry	Inserts; reinforcing steel Waterproofing; shoring of openings
Structural steel	Field painting touch-up
Miscellaneous metal	Installation; ferrous metal
Sheet metal	Louvers; skylights; closures
Roofing	Flashings; wood nailers
Metal doors/frames	Weatherstripping; glazing
Ceramic tile	Lath and scratch coat
Lath & plaster	Caulking; access panels
Painting	Sandblasting
Mechanical	Formed concrete work; dewatering Cutting/patching over trenches
Electrical	Demolition; misc. metal items

Customs-of-the-trades do not only delegate certain items of work to specific trades, but they also set loose limits to the quality of workmanship. Only by knowing what standards are generally accepted by A & E's, inspectors, government agents, and owners can the estimator designate appropriate man-hours. As a rule, unless specifications decribe required standards of workmanship, estimators assume and figure the cost to accomplish customs-of-the-trades quality. Even so, standards of quality vary with the owner-group satisfaction level, inspector's personal desires, and with the prestige of the project—university building, for instance, compared to a storage warehouse. Even when specs do not qualify, estimators tend to figure higher costs for construction work items which will be seen by many people, or by critical people, than items which are inconspicuous; street side compared to alley; public park compared to industrial center, etc. This is a matter of judgment and can be controversial.

Part 2 Working Techniques

2.1 General order of procedure

The sections in this Part 2 are arranged in the order of procedure of a typical estimate; the steps are approximately as follows:

1. Terminology, abbreviations, weights, measures, useful information, mathematics, the working tools of the estimator, are discussed (Sections 2.2 through 2.12).
2. A new project is analyzed for desirability, accepted or rejected (Section 2.13).
3. The time to do the estimating is scheduled (Section 2.14).
4. Alternate bids and bid schedules are studied (Section 2.15).
5. Sitework items, demolition, earthwork, concrete, are partially taken off; the site is visited and examined (Section 2.16).
6. The specs are summarized (Section 2.17).
7. General conditions and the progress schedule are partially completed (Section 2.18 & 2.19).
8. Demolition is completed and priced out (Section 2.20).
9. Concrete, forms and structural excavation are taken off and priced out (Sections 2.21 & 2.22).
10. Rough carpentry is taken off and priced out (Section 2.23).
11. Finish carpentry is taken off and priced out (Section 2.24).
12. Miscellaneous trades are taken off and priced out (Section 2.25).
13. Fringe benfits are added to the labor estimates (Section 2.27).

When these steps have been taken, the estimating work for the general contractor's portion is substantially completed. All that remains is the assembling of the sub-bids, and the final marking up.

2.2 Terminology

Certain words and phrases are popular in the construction industry. Outside of this field, some of them would border on the ungrammatical; but in the construction context they are natural to the ear. A book on construction which avoids such terms as "hashing-over" and "plug-in figures" would seem stilted. Since *construction language* is used in this book, the following definitions are given for the benefit of those readers who may need clarification.

Activity — any portion of the construction work, regardless of size; for instance, window cleaning, installing door frames, or the entire electrical trade.

A & E — the designing architect and engineer. The term applies even if there is only one or the other.

Addendum — a written change made by the A & E to the drawings and specs *prior* to bidding.

Allowance — a money amount arbitrarily allotted to an item in lieu of an estimated amount.

Arithmetic — see math.

Back-up figures — the detailed calculations which produce final unit prices.

Break down or breakdown — verb or noun regarding the separation of a whole (project) into its parts.

Ball-park-figure — rough estimate to find the "range" of cost. See range defined.

Budget — a rough estimate of a subtrade or item which is unfamiliar to the estimator.

Calcs — engineer's formal mathematical proofs of stresses, etc.

Calculate — verb, to do mathematical work, seeking a numerical value.

Calculation — noun, the product or result of math work.

Category — one of the divisions of the breakdown, smaller than trade (see Section 1.7—Divisions of the Estimate).

Change order — a formal change in the drawings and specs *after* the bidding.

Computation — noun, same as calculation, except more formal.

Compute — verb, same as calculate, but more formal.

Contingency — unknown or unexpected cause for extra cost.

Cut — reduce a cost item.

Demo — short for demolition.

Element — smallest division of the breakdown for estimating purposes.

Estimate — verb and noun, a time/cost prediction; the act of preparing an estimate; the estimate, itself.

Extension — the product of a quantity multiplied by a unit price; the completion of a mathematical equation.

Extra — work added to a project after the contract is let.

Field — two meanings: (1) Occupation, such as a trade, profession, or specialty; (2) The actual construction project site.

Figures — numbers, digits, symbols for quantities, cost values.

Figuring — calculating, computing, working with numbers.

Firm — promise to do work; guaranteed price.

Fringe benefits — money payed by employer to various agencies on behalf of an employee, such as pension, insurance, etc.

Grade — Three meanings: (1) the elevation of earth; (2) verb, to move earth; (3) slope, usually in degrees.

Hashing-over — discussing, debating, revising estimates.

Item — a subdivision of the breakdown, smaller than a category, but larger than an element.

Job — two meanings: (1) same as construction project; (2) any work.

Left-on-the-table — the dollar difference between the low bid and the next to low bid.

Let or sublet — issue a contract for a portion of a project.

Loose estimate — one which allows for contingencies—a "safe" or high estimate.

Lump (sum) — an item or category priced as a whole, rather than broken down into its elements.

Math — mathematics, calculating, computing, figuring, arithmetic.

Markup — contractor's overhead and profit.

Main division — general contractor's work (I), or subcontractor's work (II).

Over-run— the amount by which an item costs more than was estimated.

Pick-up — the amount by which the estimate for an item is higher than the actual cost; saving, underrun.

Plans — drawings in *plan* view. The term "drawings" is all inclusive, including plans, elevations and details.

Plug-in — a temporary figure to use in an estimate until a more dependable one is obtained or estimated.

Precon — contraction for preconstruction meeting.

Prices — dollar values before actual cost is proved.

Price out — verb, the activity of applying dollar values to the items in a take-off.

Project — a yet-to-be-built structure; construction job.

Preliminary estimate — rough estimate made in an early stage of the designing, or in anticipation of a firm bid.

Quantities — list of building materials.

Quantity survey — same as take-off, but more formal.

Quote — verb, to offer a guaranteed price.

Quotation — the price as quoted.

Range — difference between high and low estimates and/or bids.

Rebar — short for reinforcing steel.

Rough estimate — an estimate made with insufficient detail, or time, or information.

Slack — the amount of contingency, or "looseness" in the estimate.

Spec — short for specification; plural, specs.

Spread — same as range.

Spread sheet — large, wide sheet with many columns, used in estimating and/or bidding.

Sub — subcontractor.

Sub bid — bid offered by subcontractor.

Sublet — issue a contract to a subcontractor.

Super — short for superintendent.

Take off — verb, the activity of making lists of materials from the drawings and specs.

Take off — noun, the actual material lists.

Tender — a formal offer of a bid.

Tight — opposite of loose—an estimate that does not allow for contingencies, has no slack.

Trade— a subdivision of a breakdown, composed of workmen specializing in a particular skill.

Under-run — opposite of over-runs; same as pick-up; the amount that actual cost is less than estimated.

Unit costs — the actual unit costs, when work is completed.

Unit prices — the estimated, or quoted prices, before work is completed.

worksheet — paper on which the calculations supporting the estimate are recorded.

2.3 Abbreviations

The following abbreviations are typical in the construction industry, and many of them are used in this book. Periods are deliberately omitted, except in cases where they are needed for clarity.

AC	— asphalt concrete	approx	— approximate
A/C	— air conditioning	A&E	— architect/engineer
al	— aluminum	arch.	— architect
alt	— alternate	bbl	— barrel
AB	— anchor bolt	bd	— board
&	— and	blk	— block

45

blkg	— blocking	grd	— grade
bldg	— building	gyp	— gypsum
bm	— beam	hdwre	— hardware
B.M.	— bench mark	horz	— horizontal
bot	— bottom	ht	— height
cal	— calendar	hp	— horsepower
carp	— carpenter	hrs	— hours
cf	— cubic foot	hyd	— hydraulic
cfm	— cubic ft/min	ID	— inside diameter
chr	— crew-hour	in	— inch
CI	— cast iron	ins	— insurance
c.in	— cubic inch	insul	— insulation
CIP	— cast-in-place	int	— interior
cir	— circle	irrig	— irrigation
circum	— circumference	jst	— joist
CL	— center line	jt	— joint
clr	— clear	lab	— labor
col	— column	lb	— pound
conc	— concrete	lam	— laminated
CB	— concrete block	ld	— load
const	— construction	lf	— linear foot
cont	— continuous	lg	— long
contr	— contractor	lin	— linear
cov'g	— covering	LS	— lump sum
ctr	— center	lt	— light
Cwt	— hundred-weight	mach	— machine
cy	— cubic yard	max	— maximum
deg	— degree	MB	— machine bolt
demo	— demolition	Mbf	— thousand bd ft
DF	— Douglas Fir	meas	— measure
det	— detail	mech	— mechanical
diam	— diameter	membr	— membrane
dim	— dimension	met	— metal
disp	— disposal	min	— minute
dp	— deep	min'm	— minimum
dr	— door	MH	— man-hour
drg	— drawing	misc	— miscellaneous
ea	— each	mt'l	— material
elev	— elevation	NIC	— not in contract
elec	— electrical	NTS	— not to scale
eng'r	— engineer	No	— number
equip	— equipment	o/c	— on center
etc	— etcetera	OD	— outside diam
exc	— excavation	opp	— opposite
exist'g	— existing	PC	— portland cement
exp	— expansion	%	— percent
ext	— exterior	perim	— perimeter
FB	— fringe benefits	plmbg	— plumbing
fin	— finish	ptg	— painting
flr	— floor	plt	— plate
ft	— foot	pt'n	— partition
ftg	— footing	prec	— precast
found	— foundation	prestr	— prestressed
ga	— gauge	r	— radius
gal	— gallon	ref	— reference
galv	— galvanize	reinf	— reinforcing
gen	— general	req'd	— required

Rwd	— Redwood	
sec	— second	
sect	— section	
sf	— square foot/square feet	
sht	— sheet	
s.in	— square inch	
sim	— similar	
specs	— specifications	
struc	— structural	
std	— standard	
stl	— steel	
sym	— symmetrical	

temp	— temperature
tot	— total
thk	— thick
typ	— typical
UON	— unless otherwise noted
wp	— waterproof
wdw	— window
ws	— waterstop
wt	— weight
yd	— yard
yr	— year

2.4 Weights, measures and useful information

Engineering and science textbooks give tables of weights and measures most of which are superfluous to the estimator. The following list includes most of the data he needs for typical projects:

1 mile	=	5,280 linear feet
1 square mile	=	640 acres
1 acre	=	43,560 square feet
1 square	=	100 square feet
1 square foot	=	144 square inches
1 square yard	=	9 square feet
1 cubic yard	=	27 cubic feet
1 cubic yard	=	150 shovels full (about)
1 cubic foot	=	1,728 cubic inches
1 cubic foot	=	7.5 gallons
1 gallon	=	231 cubic inches
1 Cwt	=	100 pounds
1 ton	=	2,000 pounds
1 barrel	=	31.5 gallons

Figure 2.4a

Cumulative Calendar Days

	Months	Days in Month	Accumulated Days
1	January	31	31
2	February	28	59
3	March	31	90
4	April	30	120
5	May	31	151
6	June	30	181
7	July	31	212
8	August	31	243
9	September	30	273
10	October	31	304
11	November	30	334
12	December	31	365

Figure 2.4b

Quick Calculation of Steel Weight

1 cubic inch	=	.283 pound
1 linear foot	=	3.4 pound (1" x 1")
1 square foot	=	40.8 pound (1" thick)

Figure 2.4c

Formula for Areas and Volumes

Area of triangle	=	length of base X ½ altitude
Area of a circle	=	3.1416 X radius squared
Perimeter of a circle	=	diam X 3.1416
Volume of a cylinder	=	radius squared X 3.1416 X height

Figure 2.4d

Formula for Estimating Cost of Form Material

Where:
sf = square feet of contact surface to be formed
k = board feet required per sf (Figure 2.4f)
u = number of times forms are to be re-used
p = average price per sf for lumber, plywood & hdwe
w = % waste

Example: 5,000 sf of forms; 3.6 bf/sf; $0.50/bf; 3 reuses
30% waste (1.30 multiplication factor)

Formula:

$$\frac{k\,w\,p}{u} \quad \text{and} \quad \frac{3.6 \times 1.30 \times .50}{3} = \$0.78/sf$$

Figure 2.4e

Weights of Earthen Materials

	lbs/cy	tons/cy
Special fill I	3,500	1.75
Special fill II	2,970	1.49
Loam (average soil)	2,750	1.38
Gravel/cr. rock	2,700	1.35
Clay, dry	2,650	1.33
Sand and clay	2,600	1.30
Sand, moist	2,500	1.25
Sand, dry	2,430	1.21

Figure 2.4f

**Board Feet of Lumber (including plywood) Required
For One Square Foot of Different Types of Forms**

	BF/SF
Slab edges 4"	1.00
Slab edges 6"	1.33
Slab edges 8"	1.67
Slab edges 10"	2.00
Footing sides (continuous)	2.25
Footing sides (spot column)	2.50
Foundation walls to 2' high	2.50
Foundation walls to 4' high	2.67
Walls 4' to 8' high	2.80
Walls 8' to 12' high	3.00
Walls 12' to 16' high	3.33
Walls 16' to 20' high	3.67
Beams w/wood shores	3.25
Spandrel beams w/wood shores	4.00
Columns	2.70
Suspended slabs w/wood shores 4"	3.60
Suspended slabs w/wood shores 6"	3.85
Suspended slabs w/wood shores 8"	4.10
Suspended stairs	7.00

Figure 2.4g

The following unit price factors for form hardware are for use in the examples given in this book. They include nails, snapties, spreaders, etc. They apply to the gross area of forms and are not affected by reuses of material.

Form Hardware Unit Costs

	Price/sf
Footings & slab edges	$.025
Foundations	.08
Walls	.12
Beams	.09
Columns	.20
Suspended slabs	.035
Stairs	.07
Misc.	.05

Figure 2.4h

Crew-Hour Costs for Carpenters & Laborers
(Foreman Pay is Included for Each Six-Man Crew)

		Laborers @ 15/hr								
		1	2	3	4	5	6	7	8	9
C	1	34	49	64	79	94	110	125	140	155
A	2	52	67	82	97	113	128	143	158	173
R	3	70	85	100	116	131	146	161	176	191
P	4	88	103	119	134	144	164	179	194	210
E @	5	106	122	137	152	167	182	197	213	228
N 18/hr	6	125	140	155	170	185	200	216	231	246
T	7	143	158	173	188	203	219	234	249	264
E	8	161	176	191	206	222	137	252	267	282
R	9	179	194	209	225	240	255	270	285	300
S	10	197	213	228	243	258	273	288	303	319

Figure 2.4i

2.5 Basic mathematics

Customarily and traditionally, decimal fractions of a foot are used for speed and simplicity. This means that feet and twelfths of a foot must be converted to feet and *tenths* of a foot. In most cases two-place decimals are sufficient. With practice the estimator becomes able to convert twelfths to tenths automatically. For example, he actually reads and writes down the figure 7'-8" as 7.67'.

Decimal Fractions of a Foot

Inches	3 Place Decimal	Use
1	.083	.08
2	.167	.17
3	.250	.25
4	.333	.34
5	.417	.42
6	.500	.50
7	.583	.58
8	.667	.67
9	.750	.75
10	.833	.83
11	.917	.92
12	1.000	1.00

Figure 2.5a

Example #1:

$$3'\text{-}7'' \times 12'\text{-}3'' = 3.58 \times 12.25$$

Example #2:

$$3'\text{-}7\frac{1}{2}'' \times 12'\text{-}3\frac{1}{4}'' = 3.62 \times 12.27$$

Explanation:
The decimal for 1", which is .08 may be further fractionalized as follows: $\frac{1}{8}$" = .01; $\frac{1}{4}$" = .02; $\frac{3}{8}$" = .03; $\frac{1}{2}$" = .04; $\frac{5}{8}$" = .05; $\frac{3}{4}$" = .06; $\frac{7}{8}$" = .07.

Figure 2.5b

The most often used formula in estimating is the
law of proportions:
a:b :: y:x (a is to b as y is to x)

Example:

If it is known that 30 doors of a certain kind actually cost $540 to install, then how much will 65 of the same kind of doors cost to install?

Solution:

$$30 : 540 :: 65: x \text{ (30 is to 540 as 65 is to x)}$$
$$30x = 540 \times 65$$
(product of extremes = product of means)
$$x = \frac{540 \times 65}{30} = \$1,170$$

This formula is hidden within the more commonly used unit pricing method, which would deal with the example problem above this way:

$$540/30 = 18/door, \text{ and } 65 \times 18 = 1,170.$$

This is one of the reasons for unit pricing in estimating—it simplifies the mathematical work.

In the use of percentages, multiplying and dividing are easier than adding and subtracting.

Example:

Increase and decrease 500 by 30%; compare methods.

Addition Method **(2 operations)**	**Multiplication Method** **(1 operation)**
$500 \times .30 = 150$ $500 + 150 = 650$	$500 \times 1.30 = 650$
Subtraction Method **(2 operations)**	**Multiplication Method** **(mental subtraction &** **1 operation)**
$500 \times .30 = 150$ $500 - 150 = 350$	$500 \times .70 = 350$

Example:

What is the percentage by which 500 has been raised to 650?

Subtract/Divide	**Divide Only**
$650 - 500 = 150$ $150 \div 500 = .30 = 30\%$	$650 \div 500 = 1.3 = 30\%$

Example:

by what percentage has 500 been reduced to 350?

Subtract/Divide	**Divide**
$500 - 350 = 150$ $150 \div 500 = .30 = 30\%$	$350 \div 500 = .70$ (mentally, $1.00 - .70 = .30 = 30\%$)

2.6
Unit quantities costs and extensions

The following units of quantities and their abbreviations are used throughout this book. With minor variations they are typical of those used throughout the construction industry.

Quantity	Each	Linear	Square	Cube	Other
inch	in	l. in	sq. in	c. in	"
foot	ft	lf	sf	cf	'
yard	yd	l. yd	sy	cy	
each	ea				1
board feet				bf	
thousand bf				Mbf	
pound	lb				#
hundred pound					Cwt
ton	t				
barrel				bbl	
square (100 sf)			sq		
acre			acre		

Figure 2.6a

Some typical materials, work operations and the units into which they are customarily divided for estimating purposes are:

machine excavation	cy, or equip hrs
hand excavation	cy
fine grading earth	sf
backfill	cy
sand, gravel, fill soil	cy or t
disposal (haul away excess)	cy or t
drilling	lf
conc forms-slab edges	lf
walls & typical	sf
concrete	cy
cement for conc	sack or bbl
lumber	bf or Mbf
plywood & gyp bd	sf
wood trim	lf
nails, bolts, hardware	lb or ea
doors, windows, etc.	ea
liquid materials	gal
carpeting	sy
roofing	sq

Seldom is it worthwhile to use more than three decimal places. Since an estimate is an "educated guess" the mathematics need only be commensurate. Here are some general guidelines:

1. When quantities are large they may be rounded off to the nearest hundred, thus:

   221,132 sf - round off to 221,100 sf
   221,958 sf - round off to 222,000 sf

2. When quantities are small they may be rounded off to the nearest ten, thus:

   881 sf - round off to 880 sf
   889 sf - round off to 890 sf

3. When quantities are small, round off the unit prices, thus:

$$15 \text{ cy} \times 32.13/\text{cy} - \text{round off to } 15 \text{ cy} \times 32/\text{cy}$$

The difference is trivial, but the time and effort saved is worthwhile.

4. When quantities are very large and unit costs small, carry out decimals to three or four places, thus:

$$220{,}000 \text{ square} \times .02/\text{sf}$$

The decimal place, if carried out to .024 makes the worthwhile difference of 880 square feet. As another example: 635,000 square feet extended from .02 to .0244 makes the important difference of 2,794 square feet.

5. Avoid using the symbol $, where it is clearly understood that figures are monetary.

6. Delete all fractions (decimal places) in both the quantities and extensions. The only fractions required are in the unit prices—and in many cases they may be omitted.

Example:	Quantity	Unit Cost	Extension
	620 cy	4.20	2,604

This method not only saves time and space, but reduces the chance of mistakes due to misplaced decimal points (see Section 1.13—Mistakes, Errors and Slips).

It is important to remember that unit prices are properly the *end* (result) of a set of calculations, not the beginning; their main purpose is pivotal use in calculations, and for checking and comparing. They are only a convenience as tools for accomplishing the real objects— total dollar estimates.

2.7 Man-hours and pricing out

Although unit prices are a convenience in estimating, the cost of labor is really a matter of man-hours and pay scales. The most accurate estimating avoids the handy, recorded unit costs (except for reference) and uses crew-hours and appropriate pay scales, thus:

When cost data would advise: 3,000 square feet at 2.70/sf = 8,100, figure instead:

carpenter foreman	80 hrs @	19.38 =	1,550
carpenters	240 hrs @	18.13 =	4,351
laborers	160 hrs @	15.10 =	2,416
	480 hrs		8,317

$$\frac{8317}{3000} = 2.77/\text{sf}$$

Because of the time and effort to do each and every item this way, it would be prohibitive; but it would be unsafe for an estimator to depend entirely upon pre-recorded unit costs.

There is a way that both unit costs and man-hours can be used to check, correct and/or reinforce each other. In the above example the carpenter trade is about the average of the pay scales, so:

$$480 \text{ hours at } 18.13 = 8{,}784; \frac{8{,}784}{3{,}000} = 2.93$$

which is a bit higher than suggested by the cost data; but in estimating practice it is a good idea to favor the higher of two prices.

A good routine in estimating is to price out every item possible both ways at the same time, so that there is a constant cross-checking, thus:

					From Records	Use
set screeds	10,000 sf	$\dfrac{3 \text{ men } 16 \text{ hrs @ } 18.13}{10,000}$	=	.087	.075	.08
place conc	126 cy	$\dfrac{6 \text{ men } 8 \text{ hrs @ } 15.10}{126}$	=	5.75	5.00	5.50
Trowel fin	10,000 sf	$\dfrac{17 \text{ men } 8 \text{ hrs @ } 17.29}{10,000}$	=	.235	.25	.24
Membr cure	10,000 sf	$\dfrac{3 \text{ men } 8 \text{ hrs @ } 15.10}{10,000}$	=	.036	.04	.04

Since men are paid for half-days, or whole (8 hour) days, figuring labor by man-hours, rather than unit prices, builds into the estimate a slight contingency and updates the costs to meet inflation.

Some estimators price out in one group all structural excavation; in another group, all concrete forming; and in still another group, all concrete placing. Separating them this way requires that they be taken out of context. Instead, a lot can be said in favor of keeping all the elements grouped in close association with their "family", thus:

> footings— layout
> mach excavation
> forming
> concrete
> curing
> backfilling
> disposal of (excess dirt)
>
> (see Section 2.12—Quantity Survey)

Five reasons for keeping the elements in context are:
1. To avoid errors of omission
2. To increase accuracy because of individual pricing out
3. For ease of making revisions
4. For useful reference
5. To provide a natural and understandable format for supervisors

The general relationship between production and quality is indicated in Figure 2.7a. Cost is represented by the numbers 1 through 5; the quickest cost rise is from A to E; no change in cost occurs along the dotted lines C to G, H to B and F to D.

The average unit costs of labor, as found in cost records and cost reference books are at the central point of the graph, and it is the estimator's job to judge variations of particular items from the average.

Example #1:
Average cost of labor to install hardwood wall paneling is .40 per square foot; estimator judges quality requirements 25% higher and production 20% lower than average.

Solution: both conditions are price-increasing, therefore—
$$.40 + (25\% + 20\%) = .40 + 45\% =$$
$$.40 \times 1.45 = .58 \text{ per sf}$$

Example #2:
Average cost of architectural concrete finish is .55 per square foot; quality 10% greater and production 20% greater than average.
Solution: production increases, quality decreases—
$$.55 - (10\% - 20\%) = .55 - 10\% =$$
$$.55 \times .90 = .495 \text{ per sf}$$

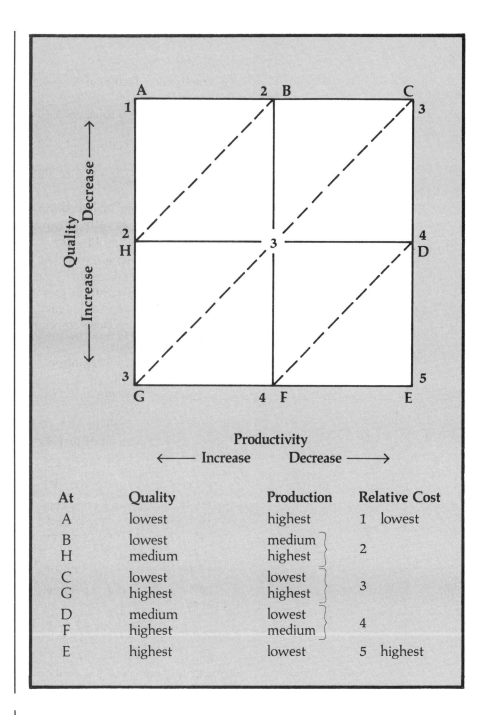

Figure 2.7a

At	Quality	Production	Relative Cost	
A	lowest	highest	1	lowest
B	lowest	medium	2	
H	medium	highest		
C	lowest	lowest	3	
G	highest	highest		
D	medium	lowest	4	
F	highest	medium		
E	highest	lowest	5	highest

2.8 Labor

The next four sections deal with the use of the price-out sheet (Figure 1.10e).

In the construction business, labor means anything for which people are paid directly (except professional services) and personally, whether it be for physical, or mental work, or a combination.

What distinguishes labor as such is its obligation upon the employer to pay not only wages, but also payroll taxes, insurances and fringe benefits.

Figuring the cost of labor is the most difficult and inexact portion of an estimate. An item of work may be composed of many elements and the cost may include many indirect and time-consuming motions. The cost to hang doors, for instance, includes receiving, unloading from trucks, storing, protecting, distributing, measuring, fitting, routing, drilling, hanging, installing hardware and adjusting—yet, it is customary to price the labor as a lump sum price per door.

The cost of labor divides into two parts:
1. The gross amount of the paycheck
2. The payroll taxes, insurances and fringe benefits paid by the employer to various agencies on behalf of the employee. For simplicity all of this is referred to in this book as "fringe benefits", or FB.

As a rule, contractors are bound by the project specifications to pay predetermined wages, thus bidders are placed on a more-or-less equal level for fairness in competition. The main trades employed directly by the general contractor are carpenters, cement finishers, and general construction laborers. Wages vary periodically. For current pay scales, the reader is referred to R.S. Means *Building Construction Cost Data*. The following, typical pay rates, are the basis of all estimating examples used in this book:

Carpenters	18.13 per hour
Cement masons	17.29 per hour
Laborers	15.10 per hour
Carpenter foreman	19.38 per hour
Cement mason foreman	18.29 per hour
Laborer foreman	15.85 per hour

These are at payroll level (the gross amount of pay checks); for fringe benefits (FB) see Section 2.27—Fringe Benefits/Payroll Taxes.

The general contractor will occasionally hire other trades, such as steel workers, welders, equipment operators, etc., and pay the appropriate rates. Equipment operators are usually covered in the equipment rental unit prices (see Section 2.10—Equipment).

Labor may be priced out in one, or in a combination of the following ways:

1. Simple unit cost, such as:

    2500 sf @ .21 = 525

2. Simple production/man-hour equation, such as:

    2500 sf @ 50sf/hr = 50hr @ 15.10 = 755

3. A lump sum, with reference to its detailed calculations on a separate worksheet, thus:

    2500 sf (see worksheet) @ ls = 525

The third case is used for items which require more explanation, and perhaps sketches, than is practical to show on the price-out sheet.

When a unit price, such as .21/square foot is named, it is important to know that it is at payroll level, automatically includes the appropriate crew makeup, but *does not include fringe benefits, insurances or taxes (Fringe Benefits)*.

2.9 Materials

The column for material is a catchall for anything that is not clearly labor, equipment or subcontract. For instance, project telephone, monthly utility charges, testing and inspection costs are listed as materials only because they would be less suitable in the other three columns.

Items in the material column include:

1. Allowances, usually in the form of lump sums (such as temporary electric service-ls 1,000).

2. Extensions composed of non-material quantities and round-figure unit prices (office supplies - 18 months at 40/mo = 720).

3. Extensions composed of quantities and unit prices not normal to the type of material; for instance, paint sells by the gallon, but is usually priced by the linear or square foot (enamel two coat - 1800 square feet at .06/sf = 108).

4. Extensions composed of non-material quantities and unit prices which incorporate several different materials, such as lumber, plywood, rough hardware, and form oil - (concrete wall forms - 17,000 square feet at .60/sf = 10,200).

5. Extensions composed of quantities as directly quoted by suppliers (concrete - 280 cubic yards at 55/cy = 15,400; lumber - 36 Mbf at 468/mbf = 16,848).

Because many items are non-material and not subject to sales taxes, taxes, when applicable, should be included in each material item, rather than applied separately to the sum total of the material column. Another reason for the application of taxes to each individual item is that some materials may be exempted, under certain conditions, from taxation.

Accurate estimating of materials depends upon:

1. An accurate take-off
2. Correct specifications and descriptions
3. A realistic waste allowance
4. Firm quotations from suppliers
5. Inclusion of taxes, delivery to jobsite, unloading, storing and protecting.

The take-off is a matter of skilled plan reading and measuring (see Sections 2.12—Quantity Survey and 1.10—Forms Format and Systems of Estimating).

Correct descriptions should be found in the specifications. If they are not, the estimator may either inquire of the A & E (Section 3.8—Addenda), or he may resort to customs-of the-trades (Section 1.21—Customs-of-the-Trades).

Methods for allowing for waste are discussed in Sections 2.21—Cast-in-Place Concrete and 2.23—Rough Carpentry.

Firm quotations requested from suppliers require giving them exact lists and specifications. When that is not possible, the estimator may request from suppliers, quotations based upon assumed quantities and specifications, pending clarification from the A & E.

Sales taxes are a direct part of material costs. They are not usually included in quotations, so the estimator must add them to the net prices, but exceptions are numerous enough that their inclusion or exclusion should always be confirmed. Such costs as freight or hauling may properly be added to and included within the direct material costs, but such costs as unloading from carriers, conveying to storage and protecting are separate costs belonging to any or all of the labor, material and equipment columns.

Materials, as priced in the price-out sheets, would be on the construction site where they would need only be picked up, moved the shortest possible distance, and installed.

Since material prices constantly change, it is a good practice for the estimator to keep a handy "running list" for quick reference and up-dating. Figure 2.9a is an example, and the unit costs shown will be used in the examples of estimates and budgets given in this book. For current average prices of materials, the reader is referred to R.S. Means *Building Construction Cost Data.*

Asphalt paving mix - tank car		180.00/Ton
Brick, common backing	8" x 2-2/3" x 4"	209.00/M
face	8" x 2-2/3" x 4"	240.00/M
fire	9" x 2-1/2" x 4-1/2"	600.00/M
Carpeting - 100% Dacron Polyester		.66/sf
Caulking - Silicone cartridge	8.5 oz	6.00/ea
Concrete items - bag concrete mix	90#	2.20/ea
	60#	1.60/ea
Mortar mix/topping	60#	2.75/ea
Portland cement - plain	94#	4.95/sk
high early		5.17/sk
white		15.00/sk
Concrete block	8" x 4" x 16"	.40/ea
	8" x 8" x 16"	.74/ea
	6" x 8" x 16"	.60/ea
Curing compound	300 sf/gal	3.90/gal
Doors - SC good quality entrance door		80.00/ea
SC flush hardwood veneer,	3'-0" x 6'-8" x 1-¾"	50.00/ea
Fir paint grd	3'-0" x 6'-8" x 1-¾"	43.00/ea
HC flush Fir paint grd	3'-0" x 6'-8" x 1-⅜"	36.00/ea
HC flush hardboard	3'-0" x 6'-8" x 1-⅜"	30.00/ea
Fencing - Western Red Cedar, rough sawn		500.00/Mbf
Split rail - post and rails complete		1.50/lf
Rough Redwood, posts, rails & boards		700.00/Mbf
Fiber board	½"	.12/sf
Gypsum board	½"	.14/sf
	⅝"	.18/sf
Insulation - Mineralwool, loose	40# bag	6.00/bag
Mineralwool, batts	3" thick	.14/sf
Fiberglass, foil faced	R-19	.32/sf
	R-11	.22/sf
Lumber - Douglas Fir Construction/Standard	2 x 2	396.00/Mbf
	2 x 3	374.00/Mbf
	2 x 4	363.00/Mbf
	2 x 6	352.00/Mbf
	2 x 8	370.00/Mbf
	2 x 10	396.00/Mbf
	2 x 12	400.00/Mbf
Redwood All-heart s4s		979.00/Mbf
Common rough sawn		495.00/Mbf
Common s4s		539.00/Mbf
Pine or clear DF KD		825.00/Mbf
Nails - Common or box, bright		.40/lb
Galvanized		.47/lb

Figure 2.9a

Paint - Latex enamel, interior semi-gloss		8.80/gal
Flat		7.70/gal
Vinyl latex house paint		13.20/gal
Vinyl wood stain		5.50/pt
Acrylic stain		4.40/pt
Olympic stain		13.50/gal
Epoxy rust-stopping primer		19.80/gal
Water seal, Thompson		11.50/gal
Piping - Subdrainage asb. cement perforated	4″	1.70/lf
	6″	2.75/lf
Unperforated	4″	1.00/lf
	6″	1.90/lf
Porous concrete	4″	1.35/lf
	6″	1.85/lf
Vit. clay perforated	4″	1.60/lf
	6″	2.55/lf
Plastic, solid or perf.	3″	.52/lf
	4″	.75/lf
Piping - PVC Class 150	4″	3.30/lf
	6″	5.00/lf
Copper 'K'	1″	3.00/lf
	1½″	3.70/lf
	2″	6.00/lf
Copper 'L'	1″	2.50/lf
	1½″	3.30/lf
	2″	5.50/lf
Plastic sheeting, Polyethylene	.004″	.017/sf
	.006″	.025/sf
	.008″	.033/sf
	.010″	.037/sf
Plywood & paneling		
Plywood DF CDX	¼″	275.00/Msf
	⅜″	297.00/Msf
	½″	396.00/Msf
	⅝″	440.00/Msf
	¾″	550.00/Msf
Plyform	⅝″	480.00/Msf
	¾″	590.00/Msf
ACX	½″	520.00/Msf
Partical board	½″	.32/sf
	⅝″	.42/sf
	¾″	.49/sf
Simulated hardwood	¼″	.37/sf
Real wood veneer	¼″	.65/sf
Hardboard, medium	¼″	.25/sf
Roofing - 90# roll mineral surface		.14/sf
15# asphalt saturated felt		.045/sf
Corrugated fiberglass		.26/sf
Asphalt shingles 235#		.33/sf
Cedar shakes (medium)		.88/sf

Figure 2.9a (cont.)

Sheet metal-Gutter GI	.32/lf
Downspout	.44/lf
Gutter els for downspouts	1.50/ea
Corners	1.30/ea
Drop inlets	1.70/ea
Straps	.75/ea
St. stl sheet 20 ga	1.32/sf
Alum sheet .063" thk	1.20/sf
Alum corr. roofing .032" thk	.15/sf
Steel - Structural shapes	.41/lb
Reinforcing steel bars	.27/lb

Figure 2.9a (cont.)

2.10 Equipment

"Equipment" does not include small tools (see Section 2.18—General Conditions), but generally mechanically operated machinery and items which are either rented or owned and are being depreciated. The equipment column on the price-out sheet carries items which may not, strictly speaking, be equipment, but are related more closely to equipment than to labor, material or subcontract (example: rented scaffold).

Equipment may be classified as:

1. Bare equipment (without operator or fuel), priced as rented per hour, day, week or month. If operators and fuel are required, they may be priced out under the labor and material columns.
2. Bare equipment as above, but owned by contractor and priced at a depreciation hourly rate (whatever the contractor considers a fair return).
3. Operated equipment complete, including fuel, priced by the hour, day, week or month and entered in the equipment column exclusively.

Estimating the cost of equipment involves (1) choosing the proper type of equipment, (2) judging the time it will be used, (3) applying the correct rental rate, inclusive of operator and fuel, and (4) including the cost to move on and off the job (transporting the equipment from and back to the rental or storage yard).

The estimator's choice of equipment may not be identical to that which would later be selected by the project super, but there should be a logical similarity and a cost equivalent. The estimator's work is theoretical; ideal equipment might not be available when needed, or actual conditions affecting the type of equipment might differ from the estimator's conceptions. The choice is that type of equipment which the estimator thinks will *probably* be used.

The work to be done suggests its own equipment; for example, demolition of concrete suggests concrete saws, jackhammers, compressors, loaders and trucks; structural excavation suggests trenchers, backhoes, excavators; depositing of concrete in forms suggests buggies, conveyor belts, cranes, or concrete pumps.

When a general type of equipment is suggested by the work to be done, it is made more specific by the given conditions of location, quantity, characteristics of the material, and the required production. For instance, if certain trenching suggests a backhoe; the depth of

trench, hardness of soil, location of stockpiles and/or the need to keep
a certain number of trucks filled and moving, suggests a backhoe of a
particular production capacity.

Although the estimator might not know as much technically about
equipment as mechanics, dispatchers, salesmen, operators and project
superintendents, still he may be more skilled than any of those at
estimating equipment costs. No one is better suited than the estimator
to judge the time (production) which is essential to an accurate cost
estimate, however, the personal choice of equipment may differ.

The time to compact earth (backfill) around foundations, for instance,
varies considerably in different conditions. The estimator's knowledge,
gained from his hours of work with the current project, guides him to
the selection of a production rate at some point between the lowest
(mostly hand work) and the highest (mostly machine work).

Useful to the estimator is a table, designed and kept current by
himself, of frequently used equipment, production capacities and rental
rates, similar to the following table which will be used in the
estimating examples given in this book. For current average rental
rates, the reader is referred to R.S. Means *Building Construction Costs
Data.*

Equipment Production and Rental Rates				
Description	Capacity	Cost Per Hr Un-Operated	Cost Per Hr Operated	Suggested Max * Production
Backhoe	Under ½ cy		$ 52	20 cy/hr
	½ cy		59	30 cy/hr
	⅝ cy		63	40 cy/hr
	¾ cy		65	45 cy/hr
	1 cy		70	60 cy/hr
	1-¼ cy		80	75 cy/hr
	1-½ cy		95	90 cy/hr
	2 cy		125	130 cy/hr
Gradall G-660	⅝ cy		95	60 cy/hr
G-680	1 cy		110	100 cy/hr
G-1000	1-½ cy		125	150 cy/hr
Clamshell	½ cy		60	35 cy/hr
	¾ cy		70	45 cy/hr
	1 cy		78	50 cy/hr
	1-½ cy		110	80 cy/hr
	2 cy		150	120 cy/hr
Dragline	½ cy		52	30 cy/hr
	¾ cy		65	40 cy/hr
	1 cy		70	45 cy/hr
	1-½ cy		100	75 cy/hr
	2 cy		145	115 cy/hr
Grader small	120 hp		53	1500 sf/hr
medium	150 hp		64	2000 sf/hr
large	195 hp		70	3000 sf/hr
Scraper, self propelled	10 cy		82	120 cy/hr
	12 cy		110	144 cy/hr
	14 cy		130	168 cy/hr
	24 cy		190	288 cy/hr

Figure 2.10a

Equipment Production and Rental Rates

Description	Capacity		Cost Per Hr Un-Operated	Cost Per Hr Operated	Suggested Max * Production
Dozer		65 hp		52	40 cy/hr
	(D-3)	105 hp		60	60 cy/hr
	(D-4)	140 hp		70	80 cy/hr
	(D-5)	180 hp		80	100 cy/hr
	(D-8)	270 hp		117	160 cy/hr
		385 hp		143	200 cy/hr
	(D-9)	400 hp		160	250 cy/hr
Hydro-crane		15 ton		100	15 ton
		18 ton		103	18 ton
		25 ton		106	25 ton
		30 ton		112	30 ton
		45 ton		120	45 ton
		65 ton		145	65 ton
		75 ton		170	75 ton
Sheepfoot compactor		4' x 4'	$10		
Loader small		½ cy		45	20 cy/hr
medium		1 cy		58	50 cy/hr
large		2 cy		70	100 cy/hr
Compactor small hand oper.			$ 7		
medium walk behind			10		
large ride on			15	40	
Trencher 6" x 24" deep trench			20	40	20 cy/hr
8" x 30" deep trench			30	52	50 cy/hr
84" deep			45	72	100 cy/hr
Forklift small			18	45	
medium			23	50	
large			28	55	
Generator small	2.5 KW		4		
med.	5.0 KW		7		
large	10.0 KW		9		
Welding equip-oxy-acetyl			7		
arc-welder			9		
Trucks Tool truck w/winch			$11		*
Flat bed to 2 ton			9		
Heavy duty pickup			8		
Dump, 8 cy-2 axle				40	8 cy/ld
10 cy-3 axle				55	10 cy/ld
15 cy-5 axle (end)				60	15 cy/ld
20 cy-5 axle (bottom)				65	20 cy/ld
Stake 16 ft				63	20 ton
Water 2 axle				43	1000 gal
3 axle				52	2500 gal
6000 gal tanker				60	6000 gal
8000 gal tanker				65	8000 gal
Tractor 2 axle				45	
3 axle				50	

Figure 2.10a (cont.)

Figure 2.10a (cont.)

Equipment Production and Rental Rates				
Description	Capacity	Cost Per Hr Un-Operated	Cost Per Hr Operated	Suggested Max * Production
Trailers	40' flatbed	7		
	dbl. bottom	9		
Air compressor w/hose	100 cfm	9		
	150 cfm	10		
	250 cfm	15		
	600 cfm	25		
	750 cfm	30		
	1000 cfm	37		

* These production rates are given only for a reference base, and they represent average soil and other conditions.

Except for special cases, great perfectionism in the selection of equipment is unrealistic in estimating; however, it is a good habit for the estimator to use a systematic proceedure such as the following:

1. Establish the desired production (cubic yards per hour).
2. Find the equipment which has a maximum production capability well above the desired level.
3. If the desired production rate is greater than the capability of any one piece of equipment, choose two or more pieces (if working room on the job site permits).
4. Estimate the *probable* production.
5. Divide the quantity by the probable production rate and find the number of hours that the equipment will be used.
6. Apply the rental rate (complete with fuel and operator).
7. Add for moving on and off the job site.
8. Round out to the next higher half or whole day.

Examples will be given under the sections on excavation, concrete and carpentry.

2.11 Subcontract work

In Section 1.7—Divisions of the Estimate, it was explained that subcontract work was of secondary interest to the general construction estimator, and was outside his field of expertise. Yet, within his own everyday work, he prices out minor items which will be sublet at a future negotiated price. The fourth column in the price-out sheet is reserved for such estimates. Some of the costs entered in this column are:

1. Professional services which are not usually obtained as firm bids, such as surveying, testing of materials, inspections and engineering.
2. Whole trades which the estimator is uncertain of subletting, or which he wishes to know the approximate value of (see Section 3.1— Budgeting Subtrades), such as rough grading, millwork, drywall work.
3. Portions of trades or specialties which the estimator keeps separate from the contractor's own work on the chance they may be sublet, such as concrete curbs and sidewalks.
4. Items which are done by specialists on a unit cost basis, such as core drilling of concrete, concrete saw cutting, sandblasting and window cleaning.

Except for the indirect subcontract items described above, the estimator's only immediate interest in subcontract work is to note the trades, or portions of trades, which may become decisive to the success of the bidding.

2.12 Quantity Survey (take-off)

Even if the material, labor and equipment unit costs are faultless, if the quantities are incorrect, the entire estimate is incorrect. An estimate may be stated this way: *If* there are 3,600 square feet of concrete trowel finish, and *if* each square foot = .25, then the total would be 3,600 square feet at .25 = 900. In a logical formulation, the quantity is the first premise. Quantity surveys are not as easily checked for errors as are unit prices and extensions. In practice, they are rarely checked, as the only good check is an independent take-off by a second person (or quantity survey team). The personnel to make a second take-off are not usually available, and when they are, the cost may be prohibitive. The usual method is that the estimator-quantity surveyor works systematically, carefully, with practiced skill, and uses a system of self-checking (see Section 1.13—Mistakes, Errors and Slips). Spot checking is better than nothing, and is discussed further in this section.

Large companies may employ persons who do nothing but take off quantities. However, for accuracy in pricing out, it is strongly preferred that the estimator does his own take-off. The conditions under which quantities are installed are most clearly visualized during the take-off work. Some items may be priced out best as they are taken off (demolition, for instance) and before the cost-effecting circumstances are forgotten.

One argument in favor of an independent take-off is that, in the process of investigating for cost-affecting conditions, the estimator may discover errors and oversights in the take-off. Nevertheless, the balance tends to favor the one man quantity surveyor-estimator method.

Quantities are usually thought of as real materials, and most of them are, but some of them are simply areas that require work, such as the finishing of concrete. The quantity surveyor knows and looks for work items which are necessary in a construction project, but which may not appear on the drawings. So the take-off is not just a list of materials, it is a list of measurements segregated in a form convenient for the application of unit prices.

The main thought of the quantity surveyor is how to separate, name and describe the quantities for one purpose, which is to be priced out. He cannot take off certain quantities without first planning the way the construction work is most likely to proceed (see Section 1.14—Selecting Methods of Construction). For instance, the quantity of earth to be excavated for an open pit depends upon working room, type of soil, necessity for ramps and shoring. A considerable amount of theoretical engineering enters into the quantity surveying work, as it does in the pricing out.

In the actual construction, one division of work may flow smoothly after another or may overlap and seem to be indivisible, but it is necessary in taking off quantities to separate the project into different classifications. Natural points of separation occur between:

1. Different measurements, such as linear feet, square feet, cubic feet.
2. Different cost levels, such as continuous footings, spot footings, stem walls, slabs.

3. Different work classifications, such as laborer, carpenter, cement finisher, welder, truck driver.
4. Different operations, such as machine excavation, hand excavation, fine grading, backfilling, form work, placing and vibrating concrete, trowel finishing.

A good method of taking off is one that keeps grouped closely all of the elements associated with an item, as in Figure 2.12a. This method of grouping may be continued throughout the project. For instance, slabs on ground may include fine grading, aggregate base course, membrane waterproofing, edge forms, setting screeds, placing concrete, finishing, curing, and expansion joints; walls may include forming, placing concrete, finishing exposed surfaces, etc.

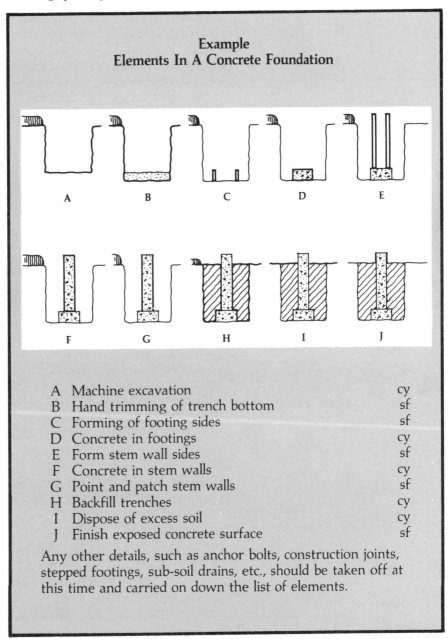

Example
Elements In A Concrete Foundation

A	Machine excavation	cy
B	Hand trimming of trench bottom	sf
C	Forming of footing sides	sf
D	Concrete in footings	cy
E	Form stem wall sides	sf
F	Concrete in stem walls	cy
G	Point and patch stem walls	sf
H	Backfill trenches	cy
I	Dispose of excess soil	cy
J	Finish exposed concrete surface	sf

Any other details, such as anchor bolts, construction joints, stepped footings, sub-soil drains, etc., should be taken off at this time and carried on down the list of elements.

Figure 2.12a

For items having few associated elements, the take-off form Figure 1.10c will serve. A large number of elements requires a form of many columns. These sample forms show how quantities are listed, with identifications, dimensions, extensions and totals.

Experience has proved the following methods of taking off quantities to be advisable:

1. Listing in an order which permits easy back-checking, as follows:

    a. Indentifying by drawing sheet, detail number, column line, floor or wing of building.
    b. Setting down dimensions in this order: length, width (or thickness) and height. A good habit is to take horizontal measurements on drawings first, and then vertical. List walls by length, thickness and height, consecutively; list slabs by length, width and thickness (vertical measurement is always last).
    c. Indicating points of the compass—North, South, East and West.
    d. Designating interior or exterior locations.
    e. Using different take-off sheets for different trades; concrete, carpentry, etc.
    f. Numbering the take-off sheets, as 1 of .....
    g. Color coding, or special numbering applied to drawings to match items on the take-off.

2. Have all extensions and additions checked for correctness by someone other than the original quantity surveyor.

3. When transferring quantities to price-out sheets, use a red pencil to check them off, thus avoiding omissions or double ups.

Spot checking to confirm the accuracy and completeness of the take-off, in lieu of having an independent take-off for comparison, may be done as follows:

1. Run a measuring wheel (map measurer) around all continuous type footings, walls, curbs etc., and compare the total reading to the sum of the linear footages on the take-off sheet; then multiply that total linear footage by an average width and depth of footing to check the total cubic feet.

2. Count the column (spot) footings and compare to the number on the take-off sheet. Multiply by the size of an average spot footing to check the total cf.

3. Take the gross square footage of a structure times the average thickness of slabs to compare the quantity of concrete in slabs.

4. Compare the total square footage of all floor and roof slabs for correctness in the take-off.

5. Find the cubic yards of concrete in a typical column and multiply by the total number of columns to compare with the take-off quantity.

6. Multiply the square feet of roof framing, wall framing and floor framing by appropriate factors to check the quantity of lumber and of plywood.

The above methods help to discover and correct *gross* errors. Small errors may pass undiscovered, but the estimator may compensate for them by habitually taking the slightly higher choices available throughout, such as doubling up at corners, taking the longer measurement (when in doubt), rounding extensions to the next higher whole number, instead of the nearest decimal place, etc.

2.13
Analyzing a project for desirability

The estimator's first task, when studying a new project, is to form a critical opinion of it before he invests a lot of time in the estimating. He seeks the answers to these questions:

1. Are there any special problems, conditions or features?
2. Would such problems be serious enough to advise rejection of the project?

Examples of conditions which could reduce the desirability of a project include:

1. Drawings and specs are too brief, requiring too many assumptions from the estimator.
2. The project is of an unusual, or technical kind with which the estimator and/or field personnel lack experience.
3. A portion of a project appears unsafe.
4. The time specified to complete the construction is inadequate or liquidated damages are too great.
5. A portion of the project specifies a material manufacturer, supplier, or subcontractor whose ability to perform the estimator questions.
6. The time to estimate and bid conflicts with another equally desirable project.
7. Special equipment is required which gives a competitor who already owns it an unfair advantage in the bidding.

Because this is a period of fault-finding, it is good practice for the estimator to be as negative as possible. Afterwards, the problems so discovered may be dealt with constructively.

After analyzing the project negatively, the estimator may look for unusually desirable features and balance them against the faults previously found. The net judgment may be a high, medium, low, or failing grade.

Once a project is rated as desirable enough to begin the estimating, analysis continues. Problems may continue to appear, but a good first analysis should preclude any later problems of major proportions. Yet, it is wise for the estimator to discuss all problems as they are discovered with the manager, superintendent, or another key person in the company.

There are other aspects of a project bearing upon its desirability that may lie outside the estimator's department—business considerations, for instance. The estimator's immediate concern is with those features which pertain to the actual cost of construction.

This first tentative analysis will be further modified as the estimating work proceeds until the final point, the consideration of which is detailed in Section 3.10—Overhead, Profit and Bond.

2.14
Scheduling the time to estimate

After it is decided that a project is desirable enough to bid, the estimator makes a rough schedule of the time available to do the estimating. Figure 2.14a is an example which works backward from the bid date; the x's indicate calendar days.

Since the estimator often works on more than one project at a time, a new project overlaps and requires meshing into his schedule. Figure 2.14b is a simplified schedule showing how a new project may possibly be fitted into the schedule of an estimating team.

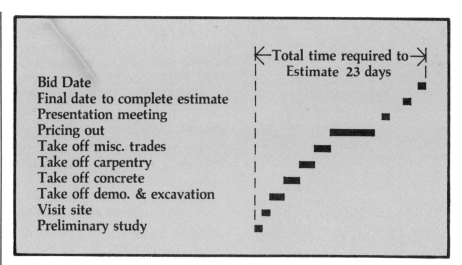

Figure 2.14a

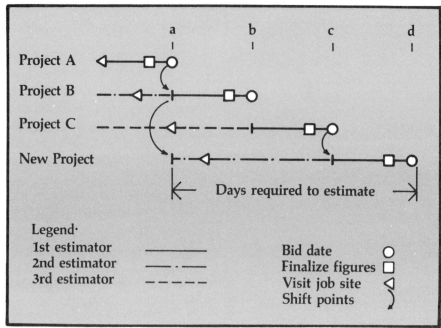

Figure 2.14b

In this example the estimating of the new project cannot start until project A bids, at which time, estimator #1 (chief estimator) takes over project B, relieving estimator #2 to start the new project. This is shift date *a*. On shift date *b* project B is completed, and estimator #1 takes over project C, relieving estimator #3 to either help with the new project or to do other follow-up duties of the department.

In this example, the chief estimator is a working estimator with two capable assistants. The schedule would be different if he supervised only; it would be different if there were but one estimator (there would not be time to do the new project, unless projects B and C were eliminated).

Estimators and projects vary widely in the time required to do the estimating; therefore, the following graphs, 2.14c and 2.14d can serve only as examples.

The hours of estimating time are *net* (uninterrupted) and include all the work normal to take-off, price-out, site investigation, presentation meeting, and bidding. They do not include the business side of bidding, nor follow-up after bidding.

Since the time to make first analyses, investigate job sites, estimate general conditions, consult with various people, and put together bids is almost the same for small as well as large projects, the curves rise rapidly at first and then gradually level off. They will never completely level off to the horizontal because, all else being equal, the greater quantity in a larger project will require more time to take off and price out.

Curve A represents simple projects (warehouses, factories, etc).

Curve B represents projects of medium complexity and cost (schools, offices, shops, etc.).

Curve C represents complex projects (hospitals, restaurants, engineering facilities, etc.).

Successful use of the chart depends upon the correctness of choice between curves A, B and C.

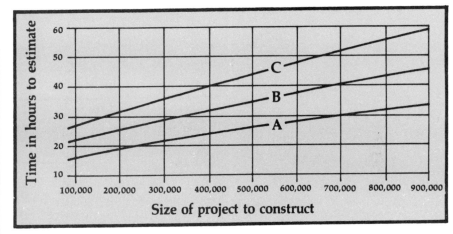

Figure 2.14c

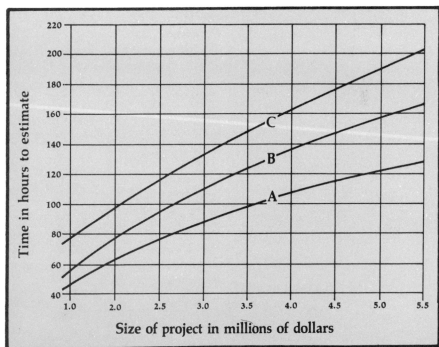

Figure 2.14d

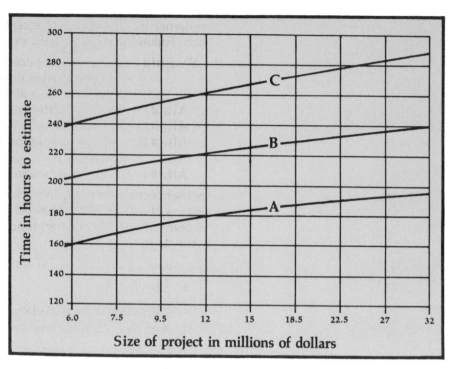

Figure 2.14e

Example#1:

A project is advertised in the cost range of two to three million dollars. The estimator judges the complexity as halfway between B and C. From the graph the time to estimate would be in the range of 70 to 122 hours, *net*. If, due to interruptions and other business, an average of only four net hours of estimating work may be accomplished in a day, then 70 + 122 ÷ 2 x 4 = 24 days will be required; this assessment may now be applied to the team schedule, as in Figure 2.14b.

Example #2:

The preliminary estimate for a project of classification A is five million dollars. The net estimating hours, from the graph, are 120.

Working days are 120/4 = 30.

Example #3:

An extremely complicated project of above C complexity, expected to cost about three million dollars, may require 160 net hours for estimating;

Working days are 160/4 = 40.

The graph and all of the above examples are predicated on a one-man quantity surveyor/estimator. The time may be reduced somewhat by dividing the work between two or more persons.

2.15 Alternate bids and bid schedule

Before starting work, the estimator should obtain at least a general impression of the bidding format because the method of taking off may be affected by the bid schedule. Portions of the project which bid separately need to be segregated on the quantity sheets. The requirements of alternate bids may also influence the site investigation; for instance, if an alternate bid calls for the demolition of a portion of an existing building, complications may occur which would call for a special way of studying the site.

A base bid usually consists of the majority of the work in a project. Alternate bids may add to or take away from the base bid work, and the owner has the option of accepting or rejecting any or all of the bid items. Following is a simplified example.

Base bid— A lump sum price for all work shown on the drawings and described in the specifications, except work under the alternates #2, 3 and 4 described below.

Alt. #1— Deduct north wing of the building, complete.

Alt. #2— Add carpeting as indicated in the finish schedule.

Alt. #3— Add parking lot on the north side of the building, complete.

Alt. #4— Add all landscaping and irrigation.

Another form of bid schedule provides for itemizing the costs of portions of the project, not as bids, but as information requested by the owner for his budgeting, funding and basis for progressive payments to the contractor.

A simplified example is as follows:

1. Site clearing and grading	$	_____
2. Site concrete work (curbs, walks, etc.)		_____
3. Asphalt cement paving		_____
4. Landscaping & irrigation		_____
5. All site utilities		_____
6. Building complete (except air conditioning)		_____
7. Air conditioning system		_____
Total bid	$	_____

Accuracy is vitally important in the first example of alternate bids, since the amount of the contract is dependent; a reasonable degree of accuracy is important in the second example, since the cash flow is dependent.

After studying the bid schedule, the estimator is ready to take off quantities, beginning with those which are important to investigate at the construction site.

2.16 Jobsite investigations

It is good practice for the estimator to study the site work portions of the drawings and specs thoroughly before visiting the site of a proposed construction project. Thus equipped with intelligent questions, he is able to make a thorough investigation in a short time and avoid the need for repeated trips.

Occasionally drawings are so well detailed that a site investigation does not affect an estimate already made from the drawings alone. But in many cases, a previously made estimate is substantially changed after a site investigation. The contractor is usually held responsible for the kind of knowledge obtainable from a visual inpsection of the site, whether it is clearly indicated on the drawings or not. Another way to state this is that the estimator has *three* sources of information, which are of approximately equal rank and mutually supporting; the drawings, the specs and the construction site.

In order to avoid omissions of important site-check items, a pre-designed check sheet, such as Figure 2.16a is recommended. A lot of miscellaneous information may be recorded under "remarks", such as the thickness and condition of pavement which is to be removed, the presence of materials, equipment, construction, etc., not shown on the drawings.

The taking of photographs may be a valuable aid in site investigations, particularly when remodeling work is to be done. Photographs aid in discussions with others who have not seen the site, add realism to the estimating process, and provide a record of conditions which, being later obliterated by new construction, may have legal value.

Site Investigation Check Sheet

Project _____ Bid Date _____

Location _____ A & E _____

1. Distance from home office _____

2. Subsistence for workmen required? yes _____ no _____

3. Railroad spur available? _____ how near? _____

4. Working room none _____ little _____ ample _____

5. Equipment rental available? _____ how near? _____

6. Labor available? _____ quantity? _____ skill? _____

7. Subcontractors available? _____

8. Source of water _____

9. Source of power _____ telephone _____

10. Need fences, barricades, lights, flagmen? _____

11. Soil conditions _____ hardness _____ wetness _____

12. Extent of clearing, grubbing, trees, etc. _____

13. Location of disposal area _____ fees _____

14. Source of import fill material _____

15. Security requirements _____

16. Source of concrete, lumber, etc. _____

17. Remarks _____

_____

_____

_____

Investigation made by _____ Date _____

Figure 2.16a

2.17
Summarizing
the
specs

It is important for the estimator to acquire an overall view (image) of the project as soon as possible in order to identify the trades which make up the two main divisions, (I) the general contractor's work, and (II) subcontractors' work (Section 1.7—Subdivisions of the Estimate).

This summary is first made exclusively from the specs. For the estimator's purpose, the simple form shown in Figure 1.10a serves the purpose well; it also identifies the trades which are involved in alternate bids. It is in this summarizing process that the alternate bids, which had first been studied lightly in Section 2.15, are thoroughly and completely understood.

A systematic way of summarizing is this:

1. Disregarding the index of the specs (until the very last), list the first section by number and name. The example in Figure 1.10a shows 1A—General Conditions. At this time disregard the alternate bids.

2. Scan each page of the first section, making mental notes of a general nature to gain familiarity with the project; then list the next section, such as 2A—Demolition.

3. Continuing to scan each page of each section and listing the sections by name, look for trades which might be described within a section, but which should be listed separately on the summary sheet.

   Example: reinforcing steel may be described in the concrete section, but both concrete and reinforcing steel should be named separately on the summary sheet so that sub-bids or estimates may be provided for each.

4. This summary of trades becomes the outline which will be used as the total display in the final, or bidding stage. Opposite each one of these trades will be a dollar value, either computed by the estimator, or as a firm sub-bid. It is important that each trade's specs section and/or paragraph number be shown so that speedy reference may be made, and so that everything in the specs can be checked off (nothing missed).

5. To prove that everything in the specs has been checked off, use a colored pencil to highlight the important words and sentences. This shading and marking provides a quick way of later finding again, for deeper study, the signficant requirements.

6. As the scanning and highlighting continue, use the Notes and Questions Sheet (Figure 1.10b) to record odds and ends for future investigations. Look also for borderline items which might be omitted by subs (see Section 1.21—Customs-of-the-Trades).

7. Compare the completed summary to the specs index. If the index lists a section that is missing from the specs, an investigation should be made to avoid an error of omission. If the specs contain a section that is not shown in the index, a study of the drawings may show if the trade is intended—or a question directed to the A & E may be in order.

8. Finally, complete the study of alternate bids (begun in Section 2.15—Alternate Bids and Bid Schedules). Indicate the trades affected as shown on the sample Specifications Summary Sheet, Figure 1.10a.

This entire process will take a concentrating hour or two to do, and will familiarize the estimator in a general way with the overall project. In addition to the written summary and the side notes for his immediate reference, the estimator will have memorized many bits of information to guide him as he works.

The summary, as described above, is not complete until all the take-off and estimating is finished, and it may be modified by information gathered from the drawings.

Necessary to the construction of a project are such indirect costs as: field office, telephone, temporary power, security fences, etc. Such costs divide naturally into two parts: *general* and *special* conditions.

General conditions are those which are more or less typical of all construction projects. Special conditions are those which are peculiar to a particular project. For brevity, all of them are called "general conditions".

There are so many possible general conditions that the estimator requires a check list, such as the following, to ensure that nothing is overlooked.

General Conditions Check List

General Conditions
    Surveying
    Layout for structures
    Office, contractor's
    Office, inspector's
    Storage shed
    Toilets
    Telephone
    Water hookup
    Water monthly charges
    Electrical pole and service
    Electrical monthly charges
    Security yard fence and gates
    Signs, project and safety
    Temporary closures, doors and windows
    Office equipment, supplies and printing
    Small tool purchase, sharpening, depreciation
    Miscellaneous equipment rental
    Oil, fuel, tires, repairs, servicing
    Pickup trucks, transportation
    First aid and fire equipment
    Dust control, noise control
    Cleanup, progressive
    Cleanup, final
    Window and fixture cleaning
    Superintendent

Special Conditions
    Travel, for administrative staff
    Travel, for workmen
    Subsistence
    Freight costs to jobsite
    Generator for electrical power
    Barricades and canopies
    Scaffolding
    Lights, flagmen, bridges over ditches
    Sewer and water connection fees
    Building and plan checking fees and permits
    Professional services and engineering fees
    Progress schedules, CPM network analyses
    Special insurances and taxes
    Equipment purchasing
    Ferry or bridge tolls
    Shoring trench or pit sides
    Dewatering
    Premium for hot or cold weather work

Premium for unskilled labor
Temporary access roads, detours, and obliterating afterwards
Overtime work
Extra drawings and specs, not supplied by A & E
Photographs
Underpinning, protecting adjacent property
Watchmen
Office clerk, time clerk
Project manager
Bonding of subcontractors
Extra foreman
Special hoisting, elevators, towers
Environmental protection
Safety provisions, railings, etc.
Install owner-furnished materials or equipment
Locating, protecting, repairing underground utilities

It is important to note that these conditions pertain exclusively to the project at hand; these are the contractor's on-site costs, and are not to include any of his general business overhead (See Section 3.10—Overhead and Profit).

General conditions are usually the first section in the specs, and therefore the first challenge to the estimator; but at such an early stage his overview of the project lacks the detail to do a complete estimate of that subdivision. Even so, the general conditions should be started immediately, carried to a logical point, then laid aside to be completed at a later point, before the bidding stage.

The conditions of most importance to the contractor are often not mentioned at all in the specs; the estimator must use his check list, his experience and his imagination to develop the coverage of these indirect costs.

Contractors differ widely in the items which they include as general conditions, and the amounts they allow for costs. Some estimators pro-rate such items as layout, tools and supervision into trades, such as concrete, earthwork, and carpentry; therefore, estimates of general conditions for the same project, made by two estimators, cannot be compared, except item by item.

An estimator, using his particular employer's values and format can construct a scale to quickly check, or anticipate the total amount of the general conditions cost of any project. The size of a project (total cost in dollars) and the length of time (complexity) for its completion are the main factors determining the cost of the general conditions. For instance, a two-million dollar simple warehouse would require less time to construct than a two-million dollar hospital building and the general conditions cost should be less.

Figure 2.18a is an example of a scale for finding the approximate cost of the general conditions for a project.

The approximate size of a project and its time to complete are usually disclosed in the bid invitation and/or the specs. Using that information and the sample scale (Figure 2.18a), a straight line drawn from the *size of project* on the right side to the *completion time* on the left side intercepts an appropriate intermediate scale identified at the top of the figure by the size of the project. The percentage thus found, multiplied by the size of the project, yields the approximate cost of the general conditions.

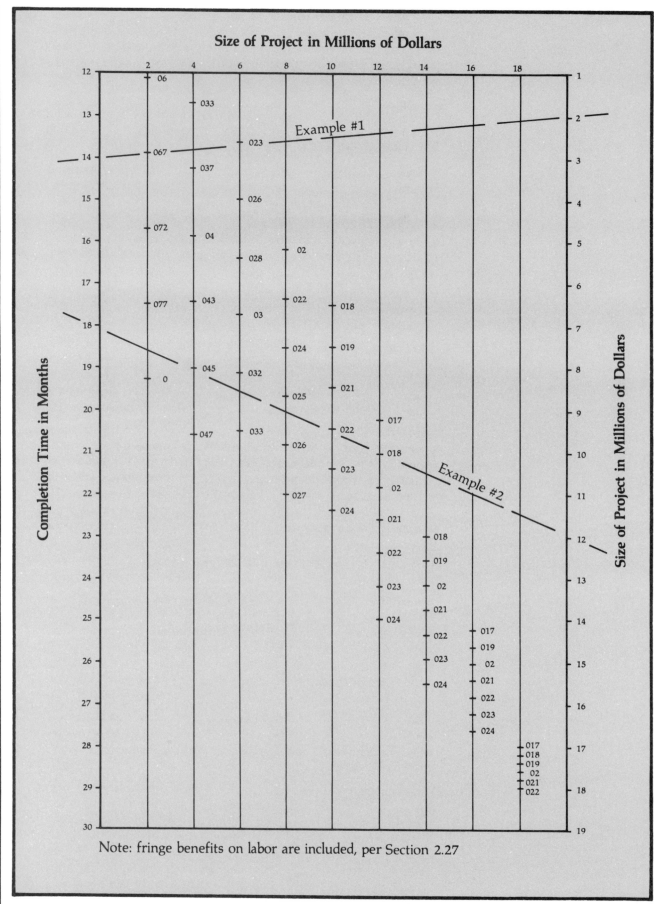

Figure 2.18a

Example #1:

A $2,000,000 project requiring 14 months to construct shows .067 (6.7%) on the graph. 2,000,000 x .067 = 134,000 approximate cost of general conditions.

Example #2:

A $12,000,000 project requiring 18 months to construct shows .018 (1.8%) on the graph in Figure 2.18a. 12,000,000 x .018 = 216,000 = the approximate cost of general conditions.

A scale such as this is not a substitute for a thoroughly itemized general conditions estimate. An indirect value is its contribution to the estimator's understanding of the three-sided relationship: complexity/time/percentage.

Such a scale is valid only for "general conditions"; *special conditions* must be added to the total produced by the use of the scale. If, in Example #2 special conditions require greater than average surveying, travel to/from jobsite, temporary access roads, all estimated at, say, 18,000, the total general and special conditions become 216,000 + 18,000 = 234,000.

Following are some typical general conditions unit prices and an example of an estimate of a $3,000,000 project requiring 18 months to complete.

Figure 2.18r is a sample of Means' Project Overhead Summary Sheet which serves both as a checklist and price-out sheet.

Contrary to what the above scale may imply, general conditions costs are not rigid and unalterable. However, there is a limit to what can be done to cut them. The most effective way is to plan methods for reducing the construction time from that allowed in the specifications (see Sections 1.14—Selecting Methods of Construction, and 2.19—Progress Schedules). Such considerations require close consultation between the estimator and the superintendent or other persons who have the power to follow through and control the economics so planned.

Figure 2.18b

Figure 2.18c

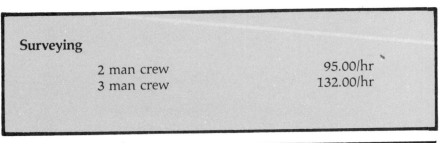

Surveying

2 man crew	95.00/hr
3 man crew	132.00/hr

Field Office

Contractor-owned trailer
Original cost, say	8,500
Maintenance, 360 x 15 yrs =	5,400
	13,900

13,900 ÷ (12 x 15) = 77.22/month

Layout Structures

Building horiz. & vert. controls before excavating

Corner batter boards	78/ea
Intersection batter boards	52/ea
Spot footing batter boards	39/ea

After excavating

Location stakes & leveling for ftgs, etc.	.13/lf
Slab elevations	.03/sf
Sidewalks	.05/sf
Curbs	.09/lf
Paving	.04/sf
Manholes & misc items	34/ea

Figure 2.18d

Storage Shed

Figure at about the same rate as the field office

say   75/mo

Figure 2.18e

Telephone

Complete cost, including installation service and long distance calls, prorated monthly   150/mo

Figure 2.18f

Portable Toilets

Complete service rental   50/mo

Figure 2.18g

Chainlink Fence, Subcontractor Install and Remove

200 lf and under	2.00/lf
201 lf to 400 lf	1.90/lf
401 lf to 800 lf	1.80/lf
801 lf to 1000 lf	1.70/lf
Over 1000 lf	1.60/lf
Add 3 strands of barbed wire	.65/lf
Add setting posts in asphaltic concrete	.30/lf
Add gate 10' x 6'	96.00/ea

Figure 2.18h

Water and Power

Water hook-up and distribution - No value will be suggested here, as conditions vary too greatly for a formula. Estimator's judgment is necessary (see Section 1.8—Judgment).

Electrical hook-up and distibution - As with water, conditions vary too greatly for a formula. Sometimes an electrical subcontractor will include temporary service in his bid, but usually it is provided at an extra charge, and in addition, the estimator makes an allowance for cords and outlets as needed at locations and floors of multi-story buildings.

Water monthly charges - vary with the size of the project, and the number of men employed.

For examples in this book, use $20/million/month.

Electrical monthly charges -

For examples in this book, use $40/million/month.

Figure 2.18i

Office Equipment and Supplies

Typewriters, calculators, file cabinets, desks, chairs, paper, pencils, drawing instruments and all the miscellaneous items required will not vary greatly in monthly cost on any project ranging in cost from one to ten million dollars. Salvage and re-use is a consideration.

For examples in this book, use $35/month

Figure 2.18j

Small Tools

Purchasing, replacement, repair and sharpening of small tools varies with the type of project. Since this cost applies to the general contractor's work only, the proportion of general to subcontract work is a rough guide to a proper allowance.

For examples in this book, use .1%/10% of general contractor work.

Figure 2.18k

Pickup Trucks, Fuel, Wear and Tear

Vehicle-miles differ with the distances and needs of projects, Assume an average number of miles per vehicle per month.

For examples in this book, use .50/mile

Figure 2.18l

Miscellaneous Rental Equipment

Some projects may require the use of general purpose equipment which does not suitably fall within a special trade. Examples are: forklifts, skiploaders, pumps cranes and hoists. Specific equipment may be estimated individually by the rental rates as suggested in Section 2.10—Equipment; otherwise, an allowance may be made based upon experience, judgment and project cost records.

For examples in this book, use .5% of the total cost of the project

Figure 2.18m

Cleanup Labor

For examples in this book, use:

	A Simple Constr.	B Medium Complexity	C Very Complex
Structures			
Progressive cleanup	2 mhr/day	4 mhr/day	6 mhr/day
Final cleanup	.06/sf	.15/sf	.24/sf
Site work			
Progressive cleanup	.03/sf	.06/sf	.09/sf
Final cleanup	.02/sf	.04/sf	.06/sf

Figure 2.18n

Project Superintendent

For examples in this book, superintendent base pay is $2/hr above journeyman carpenter, or 18.13 + 2.00 = 20.13/hr.

Figure 2.18o

Progress Schedules (CPM) Computerized

For examples in this book multiply the total amount of the anticipated bid by:

Size of Project in Dollars	A Simple Project	B Medium Complex	C Complex Project
1 to 2 million	.003	.004	.005
2 to 4 million	.0015	.002	.0025
4 to 6 million	.001	.0015	.002
6 to 10 million	.0008	.0012	.0015

Figure 2.18p

General Conditions Example of a $3,000,000 Project (18 Months)

Description	Quant.	Unit Costs			Total Costs			
		l	m/e	s	L	M/E	S	T
1. Surveying	32 hrs	—	—	132	—	—	4,224	4,224
2. Layout	ls	—	—	—	3,134	313	—	3,447
3. Temp. office	ls	—	—	—	261	1,870	—	2,131
4. storage shed	18 mo	—	75	—	—	1,350	—	1,350
5. telephone	18 mo	—	50	—	—	900	—	900
6. toilets	54 mo	—	—	50	—	—	2,700	2,700
7. fence with/2 gates	720 lf	—	—	2.07	—	—	1,490	1,490
8. Water-hookup/distribute	ls	—	60	—	360	800	—	1,160
9. monthly charge	18 mo	—	60	—	—	1,080	—	1,080
10. Power hookup/distribute	ls	—	—	—	—	—	1,800	1,800
11. monthly charge	18 mo	—	120	—	—	2,160	—	2,160
12. Office equipment and supplies	18 mo	—	35	—	630	—	—	630
13. Small tools	ls	—	—	—	—	6,000	—	6,000
14. Pickups, fuel, etc.	19,800 mi	—	0.50	—	—	9,900	—	9,900
15. Misc. rental equipment	ls	—	—	—	—	15,000	—	15,000
16. Cleanup	ls	—	—	—	15,777	3,155	—	18,932
17. Superintendent	18 mo	3,489	—	—	62,802	—	—	62,802
Subtotal					82,334	43,158	10,214	135,706
FB 46% of labor								37,874
								173,580

Note: From Fig. 2.18a the anticipated cost would be: 3,000,000 @ .06 = 180,000

Figure 2.18q

PROJECT OVERHEAD SUMMARY

MEANSCO FORM 112

PROJECT _____
LOCATION _____
ARCHITECT _____
OWNER _____

QUANTITIES BY ____ PRICES BY ____ EXTENSIONS BY ____ CHECKED BY ____
SHEET NO. ____ ESTIMATE NO. ____ DATE ____

DESCRIPTION	QUANTITY	UNIT	MATERIAL UNIT	MATERIAL TOTAL	LABOR UNIT	LABOR TOTAL	TOTAL COST UNIT	TOTAL COST TOTAL
Job Organization: Superintendent								
Accounting and bookkeeping								
Timekeeper and material clerk								
Clerical								
Shop								
Safety, watchman and first aid								
Engineering: Layout								
Quantities								
Inspection								
Shop drawings								
Drafting & extra prints								
Testing: Soil								
Materials								
Structural								
Supplies: Office								
Shop								
Utilities: Light and power								
Water								
Heating								
Equipment: Rental								
Light trucks								
Freight and hauling								
Loading, unloading, erecting, etc.								
Maintenance								
Travel Expense								
Main office personnel								
Freight and Express								
Demurrage								
Hauling, misc.								
Advertising								
Signs and Barricades								
Temporary fences								
Temporary stairs, ladders & floors								
Photos								
Page total								

R.S. MEANS CO., INC. KINGSTON, MA. 02364

DESCRIPTION	QUANTITY	UNIT	MATERIAL UNIT	MATERIAL TOTAL	LABOR UNIT	LABOR TOTAL	TOTAL COST UNIT	TOTAL COST TOTAL
Total Brought Forward								
Legal								
Medical and Hospitalization								
Field Offices								
Office furniture and equipment								
Telephones								
Heat and Light								
Temporary toilets								
Storage areas and sheds								
Permits: Building								
Misc.								
Insurance								
Bonds								
Interest								
Taxes								
Cutting and Patching & Punch list								
Winter Protection								
Temporary heat								
Snow plowing								
Thawing materials								
Temporary Roads								
Repairs to adjacent property								
Pumping								
Scaffolding								
Small Tools								
Clean up								
Contingencies								
Main Office Expense								
Special Items								
Total: Transfer to Meansco Form 110 or 115								

R.S. MEANS CO., INC. KINGSTON, MA. 02364

Figure 2.18r

82

The example of a general conditions estimate in Figure 2.18q is explained in detail as follows. More briefly, this information may be shown in the Work Sheet Form, Figure 1.10f, for the record.

1. Surveying—rough and finish staking for grading, paving, underground utilities and location's of structures is judged, by itemizing the work, to require 32 hours of a three-man crew (Figure 2.18b). Because it is done by independent contractor, though not necessarily under a formal contract, it is carried under the subcontract column.

2. Layout - Figure 2.18d

Bldg before exc - corners	6 ea @ 52.00 =	312
Bldg before exc - intersection	6 ea @ 34.00 =	204
Bldg before exc - spot footings	13 ea @ 26.00 =	338
Bldg after exc - ftg stakes	20 ea @ 13.00 =	260
Bldg after exc - slabs	32,000 sf @ .03 =	960
Site work - walks	3,600 sf @ .05 =	180
Site work - curbs	1,800 lf @ .09 =	162
Site work - paving	12,000 sf @ .04 =	480
Site work - misc	7 ea @ 34.00 =	238
Total Labor		3,134
Mat'l & equip 10% of labor		313
Total		3,447

3. Field office - Figure 2.18c

Rent/depreciation	18 mo @ 77.27 =	1,390
Move on/off - equip	8 hr @ 60.00 =	480
Total mat'l & equip		1,870
Labor-set up/dismantle	16 hr @ 16.33 =	261
Total		2,131

4. Storage shed - See Figure 2.18e.

5. Telephone - See Figure 2.18f.

6. Toilets - Allow one portable chemical unit for each million dollars of project size; in this case, 3 each @ 18 months = 54 months; See Figure 2.18g.

7. Fence w/2 gates, Figure 2.18h.

fence	720 lf @ 1.80 =	1,296
gates	2 ea @ 96.00 =	192
		1,488

$$\frac{1,488}{720} = 2.07/lf$$

8. Water - Figure 2.18i
Hookup & distribute

mat'l & equip, say		800
labor, say	20 hr @ 18.00 =	360
		1,160

9. Water - Figure 2.18i
Monthly charge - 3 (million) x 20.000 =   60/mo

10. Power - Figure 2.18i
Hookup - subcontractor, say,   1,800

11. Power - Figure 2.18i.
Monthly charge - 3 (million) x 40.00 =   120/mo

12. Office equip & supplies, Figure 2.18j   35/mo

13. Small tools, Figure 2.18k
    In this example, general contractor's work
    is expected to be 20% of the total cost, so
    $\frac{20\%}{10\%} = 2$, and 2 x .001 x 3,000,000 =   6,000

14. Pickups, fuel, etc., Figure 2.18l
    2 vehicles @ 550 miles per month each;

    2 x 550 x 18 = 19,800 miles

15. Misc. equip rental, Figure 2.18m

    3,000,000 x .005 = 15,000

16. Cleanup, Figure 2.18n
    Structures (A-simple construction)
       Progressive   756 hrs @ 14.52 = 10,977
       Final   30,000 sf @   .06   1,800
    Site work
       Progressive   60,000 sf @ .03 = 1,800
       Final   60,000 sf @ .02 = 1,200
          Total labor   15,777
       Trucks & loaders - say 20% of labor   3,155
          Total   18,932

17. Superintendent, Figure 2.18o
    The true number of working hours per month is
    $\frac{52 \times 40}{12}$ = 173.33, and 173.33 x 20.13 = 3,489/month

The example, Figure 2.18q, of a general conditions estimate agrees
reasonably with the graphically computed Example #2, thus:

Example #2   180,000
Estimate, Figure 2.18q 173,580

Both of these figures are based on *typical* conditions and average costs.
Part 4 of this book consists of a hypothetical project in which the
general conditions include some non-typical items, and shows how the
detailed estimate and the graphical result (by Figure 2.18a) are
equated.

2.19
Progress
schedules

The total estimated cost of a project is affected by the time required
to complete the construction. As indicated in Figure 2.19a, cost
increases at both ends, too little or too much time. If the contract time
is too short, the estimator must figure overtime and other production-
increasing methods. If more time is figured than is actually required,
the general conditions costs are excessive.

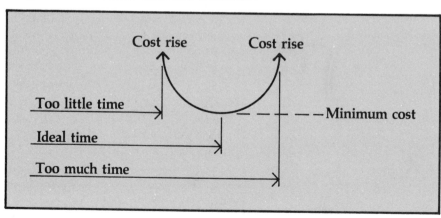

Figure 2.19a

Most specs provide a maximum time period (calendar days) for the completion of the construction, and liquidated damages (a penalty charge for each day of delay in completion). It is desirable that the estimator make his own independent judgment of the ideal time period, from the contractor's (rather than the owner's) viewpoint.

Example #1:

A project specifies 420 calendar days and $250 per day liquidated damages; but by his own formula the estimator believes the ideal time should be 450 days (30 days longer). He may:

1. Take a chance on the specified time =$0
2. Include in the estimate an amount for over-run labor cost
   Example: 320 hrs @ $24 = 7,680
3. Include an amount to pay liquidated damages.
   Example: 30 days @ $250 = 7,500
4. Estimate the actual cost of an extra month of general conditions
   Example: 7,100
5. Use a combination of the above, or split the difference, thus:
   $$\frac{0 + 7,680 + 7,500 + 7,100}{4} = 5,570$$

Note that: choice 1 is given the weight of "0" thus influencing the final, average, amount.

Example #2:

A project specifies 510 calendar days; the estimator judges 450 days (2 months less). The general conditions cost is reduced by:

$$2 @ 7,100 = 14,200$$

An experienced estimator is able to judge the approximate time required for the construction of a project, but it is preferable for him to draw a simple graph, as in Figure 2.19b, to demonstrate and prove his judgment.

In this graph the time span of each trade is a mere judgment; however, a single trade is easier to judge for time accurately than an entire project and, since the trades are not strung end to end like beads, but overlap, their *net* time is that period required to complete the entire project. This graph may be used later to construct a more detailed progress schedule for the guidance of the project super and all persons involved in the actual field construction work (see Section 3.11—Follow-Up).

Figure 2.19c is a sample of Means' Progress Schedule form, designed for this purpose.

The key to the completion time might be any of the major trades (masonry, structural steel, mechanical, electrical), but the most common is the structural shell of the building. After having estimated this part of the project, the estimator has a very good insight regarding the time to complete the footings, slabs, columns, beams, floors, roof, stairs, and so on. Most of the subtrades fall within a schedule constructed on the structural "shell".

A side benefit of budgeting the time of subtrades (see Section 3.1—Budgeting Subtrades) is the insight gained by the estimator regarding the time requirements to be allowed in the progress schedule—information which even the subcontractng specialists may not be competent to judge.

In Figure 2.19b, the schedule indicates the need for 17 months, rather than the contractural maximum of 15. Schedules for other projects may agree with, or show the need for less time than that specified.

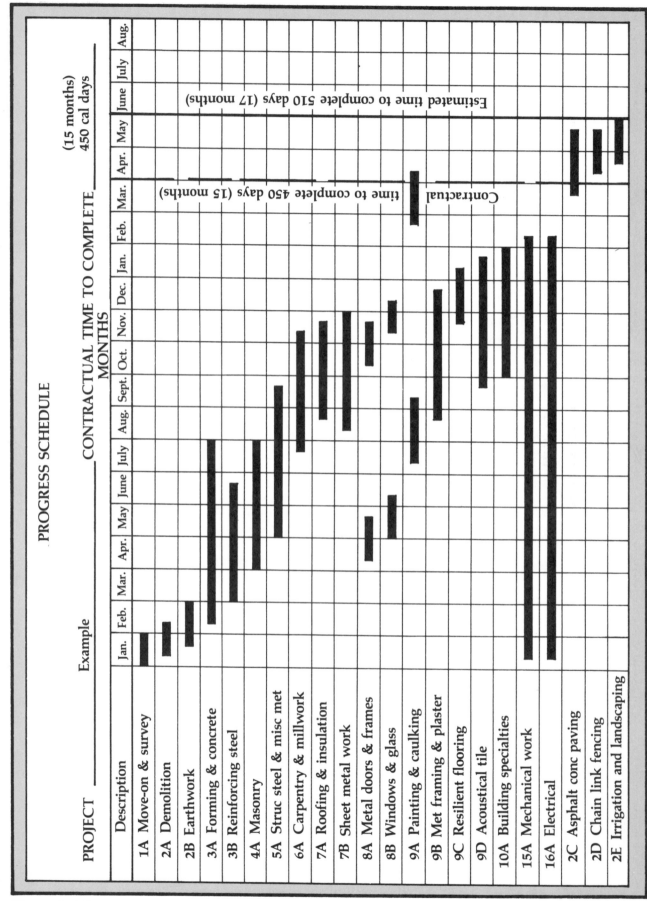

Figure 2.19b

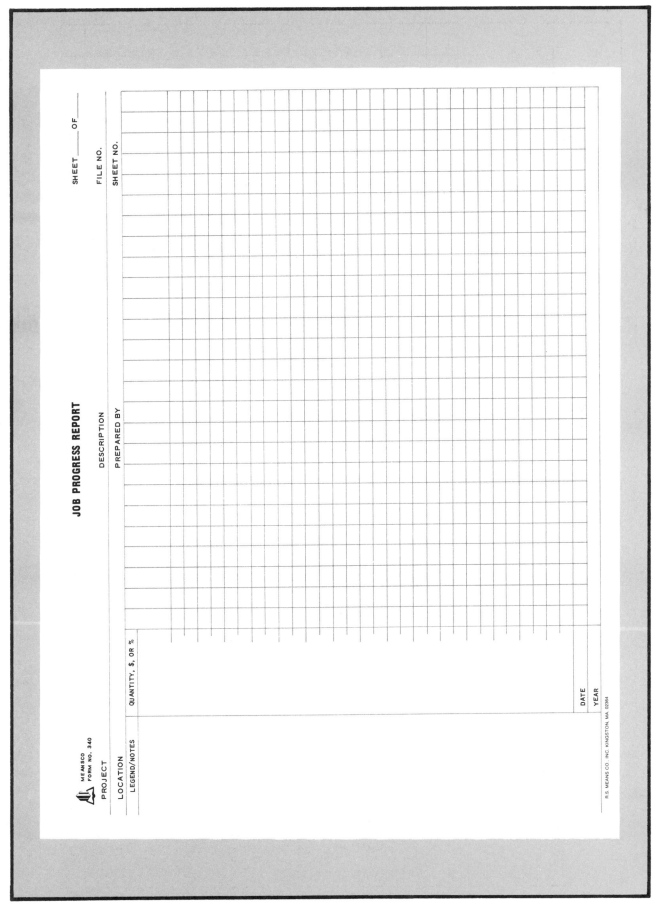

Figure 2.19c

Practice in such graphing increases the estimator's self-confidence in the correctness of his general conditions cost estimates.

The best time to construct the graph is toward the end of the estimating work, at which point the estimator most clearly visualizes the scope and relationships of all the trades.

2.20 Demolition

The first task on a project is the removal of obstacles in the way of the new construction, and the estimator usually figures this "demo" first, because of the information the take-off provides necessary to an intelligent jobsite investigation (see Section 2.16—Jobsite Investigations).

Demolition work should be taken off in three separate groups, namely:

1. Site clearing, including buildings and structures, which are to be totally removed. This group should be done in readiness for comparison to possible sub-bids, and for this reason its description is not included in this section (see Section 3.1—Budgeting Subtrades).
2. Items of site clearing demo which subcontractors do not customarily include in their bids, such as saw-cutting for paving, moving buildings, demo for remodeling, etc. (see Section 1.21—Customs-of-the-Trades).
3. Cutting and removing portions of existing structures incidental to remodeling.

It is expected that Group 1 will be sublet; Group 2 is uncertain as to subletting; Group 3 will probably not be sublet.

Typically, demolition does not adapt well to take-off sheets. Most demolition items are individual and unique. Material cost is rarely involved, so detail is not as important as in new construction. Each item as measured on the drawing may be printed directly onto the price-out sheet (see Figure 1.10e).

Unlike new construction, which is taken off in the units by which materials are purchased (board feet, pounds, cubic yards, etc.), demo is taken off in the form by which items are handled (tons, truckloads, equipment-hours, man-hours, etc.).

Speed is a leading aim in demo. New construction cannot begin until obstacles are removed; however, for the convenience of the builder or owner, in some cases particular items are postponed to later periods in the construction schedule for demolition.

Equipment is usually more economical in time and cost than hand labor; consequently, equipment is usually the largest category of cost on the price-out sheet. Most of the labor, since it consists of equipment operators, is included within the equipment unit costs (see Section 2.10—Equipment).

Direct labor is necessary for demolition of items that are small, require special care, or are inaccessible for equipment. The specs may direct that certain items be dismantled with care for salvage to the owner, or the estimator may decide that certain items have salvage value to the contractor. Here are some typical unit costs (requiring regular revision for inflation), followed by an example of a demolition price-out and explanation.

The demo estimate, Figure 2.20p, is explained as follows:

All unit prices used are for average conditions; those used in a *real* estimate would probably run higher or lower due to non-average conditions.

Saw Cutting Concrete Slabs Per Linear Foot

Depth of Cut	To 500'	500' to 1,000'	1,000' to 5,000'
½"	.50	.25	.19
1"	1.06	.53	.40
1½"	1.25	.63	.47
2"	1.56	.81	.61
2½"	1.88	.94	.71
3"	2.50	1.25	1.13
3½"	3.00	1.55	1.25
4"	3.75	1.87	1.40

Figure 2.20a

Saw Cutting Asphalt Concrete Paving Per Linear Foot

Depth of Cut	To 500'	500' to 1,000'	1,000' to 5,000'
1"	.45	.38	.30
1½"	.50	.45	.35
2"	.55	.50	.40
2½"	.60	.55	.45
3"	.67	.60	.52
3½"	.72	.65	.57
4"	.80	.72	.65

Figure 2.20b

Saw Cutting Vertical Concrete Walls Per Linear Foot

Depth of Cut	To 50'	50' to 100'	100' to 500'
⅜"	1.50	1.25	1.00
½"	1.80	1.50	1.20
⅝"	2.10	1.75	1.40
¾"	2.63	2.19	1.75
1"	3.15	2.62	2.10
1½"	5.40	4.50	3.60
2"	7.20	6.00	4.80
3"	9.00	7.50	6.00

Figure 2.20c

Core Drilling Reinforced Concrete Slabs (Downward)
Cost Per Hole Foot

Diameter of Core	Cost/Linear Feet
1"	26.00
3"	36.00
4"	46.00
6"	65.00
8"	85.00
10"	130.00
12"	163.00
14"	227.00
18"	455.00

For horizontal coring, add 50% to the above
For overhead coring, add 100% to the above

Figure 2.20d

Figure 2.20e

Drill Concrete for Anchors

Diameter of Hole	Cost/Linear Inch
½"	.82
¾"	.98
1"	1.22
1½"	1.87
2"	2.43

Figure 2.20f

Remove Concrete Slabs on Ground
Per Square Foot (Pneumatic Tools)

Slab Thickness	Unreinforced		Reinforced	
	Labor	Equip	Labor	Equip
4"	.21	.17	.30	.25
5"	.25	.23	.38	.34
6"	.30	.25	.44	.36
8"	.39	.33	.59	.49

Figure 2.20g

Remove Concrete Slabs on Ground
Per Square Foot (All Machine Work)
Unreinforced

Thickness	To 1,000 sf	1,000 sf To 5,000 sf	5,000 sf To 10,000 sf	Over 10,000 sf
4"	.42	.35	.30	.27
6"	.53	.44	.37	.33
8"	.66	.55	.47	.42
10"	.81	.68	.58	.52
12"	.94	.78	.66	.59

Figure 2.20h

Remove Concrete Curbs Per Linear Foot

	Labor	Equip
Plain curb 6" wide @ top	.52	.52
Plain curb 8" wide @ top	.65	.65
Gutter type curb	.97	.65
Roll type curb	.80	.60

Figure 2.20i

Remove Concrete Footings and Foundations
and Similar Mass Concrete

	Labor	Equip
Unreinforced concrete	30.00/cy	22.00/cy
Reinforced concrete	42.00/cy	37.00/cy

Remove Asphalt Concrete (AC) Paving				
Thickness	To 2000 sf	2000 to 5000 sf	5000 to 10,000 sf	Over 10,000 SF
1"	.07	.05	.04	.035
2"	.125	.11	.09	.08
3"	.16	.145	.13	.12
4"	.20	.185	.17	.16

Remove Trees and Stumps (Medium Hardness) Each				
Diameter of Tree	Remove Trees		Remove Stumps	
	Labor	Equip	Labor	Equip
4"	7.80	5.20	3.90	10.40
6"	13.00	9.10	6.50	15.60
8"	18.20	13.00	9.10	19.50
10"	32.50	16.00	11.70	27.30
12"	39.00	19.50	14.30	32.50
16"	52.00	22.10	16.90	36.40
18"	65.00	26.00	19.50	52.00
20"	78.00	28.60	22.10	78.00
24"	104.00	32.50	24.20	104.00
30"	130.00	52.00	32.50	150.00
36"	260.00	78.00	65.00	195.00

Remove Suspended Ceilings Including Finishes/Square Foot		
	Labor	Equip
Gypsum Board	.20	.035
Lath and Plaster	.26	.065
Acoustic tile	.33	.035

Remove Partitions Per Square Foot			
		Labor	Equip
Wood stud and plywood		.26	.02
Wood stud and gypsum board		.33	.035
Metal stud and gypsum board		.53	.035
Wood stud, lath/plaster		.38	.04
Metal stud, lath/plaster		.44	.05
Concrete block, ungrouted	8"	.55	.33
Concrete block, solid grout	8"	.82	.65
Concrete, reinforced	6"	.62	.62
Concrete, reinforced	8"	.83	.83
Concrete, reinforced	10"	1.00	1.00
Concrete, reinforced	12"	1.35	1.35

**Remove Wood Building Down to, but not Including
Slab and Foundation (All Machine, no Salvage)
Per First Floor Square Foot Area**

Roofing Materials	.13/sf
Roof sheathing and framing	.23/sf
Ceiling framing and finishes	.12/sf
Wood floor framing and finishes	.25/sf
Walls, interior and exterior	.33/sf
Miscellaneous items	.23/sf
Total	1.29/sf
Or, based on 8' ceiling	.16/cf

Figure 2.20n

Remove Miscellaneous Items

Catch basins to 4' deep	1.80/cf
Catch basins to 10' deep	2.70/cf
Manholes to 7' deep	2.27/cf
Manholes to 12' deep	3.40/cf
Chainlink fence	1.30/lf
Flooring-carpeting and mat	.13/sf
resilient	.20/sf
wood	.30/sf
ceramic tile/terrazzo	.57/sf

Figure 2.20o

Under Group 2, all prices are lumped in the material/equipment column and include all labor fringe benefits, in readiness for possible subcontractor quotations, which are customarily offered in these single, all-inclusive unit prices. Items #1 and #2 would be priced higher if saw cuts were short pieces with numerous angles, and the paving were exceptionally hard. The prices could be lower for long runs and soft paving. Items #3, #4 and #5 could be priced higher or lower in consideration of the hardness of the material, the incidence of reinforcing steel, and the numbers and spacing of holes requiring moving and setting up time.

Under Group 3, the work is broken down into labor and equipment. Cost records and reference unit prices for items of alteration work are the least reliable of all trades in the craft of estimating, and more judgment is required of the estimator. Unlike simple demo, alteration work requires considerable layout work, planning and care. Accessibility problems and choice of construction (demo) methods can cause average unit price concepts to have no relation whatever to the individual case in hand.

For explanation of fringe benefits, shown here as 46%, see Section 2.27—Fringe Benefits/Payroll Taxes.

Example of a Demolition Estimate

Description	Quant.	l	m/e	s	L	M/E	S	T
Group 1 - To be sublet . (see Section 3.1—Budgeting Subtrades)								
Group 2 - Demo work by general contractor								
1. Saw cutting AC 3" deep	850 lf	—	.60	—	—	510	—	510
2. Saw cut concrete 2" deep	480 lf	—	1.56	—	—	749	—	749
3. Core drill concrete 6" diameter	16 lf	—	65.00	—	—	1,040	—	1,040
4. Drill for anchors 1½"	140 lf	—	1.87	—	—	262	—	262
5. Drill for anchors 2"	96 lf	—	2.43	—	—	233	—	233
			Total			2,794		2,794
Group 3 - Demo work for alterations								
1. Saw cut concrete, vertical 1"	180 lf	—	2.10	—	—	378	—	378
2. Concrete slab on grade, 6", reinforced	800 sf	.44	.36	—	352	288	—	640
3. Concrete footings, reinforced	30 cy	42.00	37.00	—	1,260	1,110	—	2,370
4. Acoustic tile ceiling	2,400 sf	.33	.035	—	792	84	—	876
5. Metal stud and gypsum bd. ptn's	800 sf	.53	.035	—	424	28	—	452
6. Resilient flooring	1,200 sf	.18	.02	—	216	24	—	240
			Total		3,044	1,912		4,956
			FB 46% of labor					1,400
								6,356

Figure 2.20p

2.21 Cast-in-place concrete

Concrete, forms and structural excavation are so interrelated that it is convenient to keep them together as a single trade. Everything connected with an item of concrete work, such as footings (see Figure 2.12a) is listed on the take-off sheet, then summarized on the price-out sheet and priced out. In accordance with Section 1.7—Divisions of the Estimate, a typical format for concrete work would be:

I (Main division) General contractor work
  A. (Trade) Concrete, forms & excavation
    1. (Category) Footings, continuous
      a. (Item) Layout
      b. (Item) Machine excavation
      c. (Item) Hand excavation
      d. (Item) Formwork
        (1)(Element) Steps in footings
      e. (Item) Concrete and placing
      f. (Item) Backfilling
      g. (Item) Disposal of excess dirt
    2. (Category) Column (spot) footings
      a. (Item) Layout
      b. (Item) Machine excavation
      c. (Item) Hand excavation
      d. (Item) Formwork
      e. (Item) Concrete and placing
      f. (Item) Backfilling
      g. (Item) Disposal of excess dirt
    3. (Category) Foundation walls
      a. (Item) Forming
        (1) (Element) Construction joints
        (1) (Element) Bulkhead forms
      b. (Item) Concrete and placing
        (1) (Element) Setting anchor bolts
      c. (Item) Pointing & patching
      d. (Item) Curing
      e. (Item) Finish exposed surfaces
    4. (Category) Floor slab on ground
      a. (Item) Form edges
        (1) (Element) Layout & leveling
      b. (Item) Fine grade on subgrade
      c. (Item) Aggregate sub-base
      d. (Item) Plastic membrane
      e. (Item) Set screeds
      f. (Item) Concrete & placing
        (1) (Element) Expansion & construction joints
      g. (Item) Trowel finishing
        (1) (Element) Curing
    5. (Category) Walls
      a. (Item) Forming
        (1) (Element) Bulkhead forms
        (1) (Element) Chamfer strips
        (1) (Element) Stripping & cleaning
      b. (Item) Concrete and placing
        (1) (Element) Pointing & patching
        (1) (Element) Finishing exposed surfaces
    6. (Category) Columns
      a. (Item) Forming
        (1) (Element) Chamfer strips
        (1) (Element) stripping & cleaning

  b. (Item) Concrete, placing & hoisting
   (1) (Element) Pointing & patching
   (1) (Element) Finishing exposed surfaces
 7. (Category) Beams
  a. (Item) Forming & shoring soffits
   (1) (Element) Chamfer strips
   (1) (Element) Forming sides
   (1) (Element) Stripping & cleaning
  b. (Item) Concrete, placing & hoisting
   (1) (Element) Pointing & patching
   (1) (Element) Finishing exposed surfaces
 8. (Category) Suspended floor & roof slabs
  a. (Item) Forming & shoring
   (1) (Element) Form openings, edges, etc.
   (1) (Element) stripping & cleaning
  b. (Item) Concrete, placing & hoisting
   (1) (Element) Setting screeds
   (1) (Element) Finishing
   (1) (Element) Curing & protecting
 9. (Category) Suspended stairs
  a. (Item) Forming & stripping
  b. (Item) Concrete, placing & hoisting
   (1) (Element) Finishing
   (1) (Element) Curing
   (1) (Element) Finishing soffits, edges

The above subdivisions are typical of structural concrete work; special items and elements may be inserted as they appear in particular projects; for instance, shoring of trench sides, de-watering, waterstops, sleeves, form-outs, concrete admixtures, color, special finishes, etc.

Following are explanations and examples of the above typical trades, categories, items and elements.

Layout (leveling & squaring) is generally incorporated in the various items of work; it is a separate item only in association with foundation work where its purpose is two-fold: to guide the excavation work, and to provide the horizontal and vertical controls for the formwork.

Because it is mostly a labor item, it is figured in crew-hours (chr). The number of crew-hours may be estimated by counting building corners, intersections, column (spot) footings, lines of continuous footings, and applying a formula, such as Figure 2.21a.

Figure 2.21a

Using 2 Carpenters and 1 Laborer = 52.00 per Fig. 2.4 i,			
Corner batter boards	1-½ chr	=	78.00/ea
Intersections	1 chr	=	52.00/ea
Column (spot) footings	¾ chr	=	39.00/ea
Line of cont. footings	¼ chr/100 lf	=	.13/ lf
Other (pits, etc.)	½ chr	=	26.00/ea

Machine excavation is shown on the price-out sheet in unit prices per cubic yard, even though equipment cost is based upon hours of operation. Hours are dependent upon the quantity and character of earth to be moved, the type, the capacity and the efficiency of the equipment (see Section 2.10—Equipment).

Example:

Estimate the cost to machine excavate three-feet deep 300 cubic yards in average soil.

Assume a 1/2 cubic yard backhoe costing 60.00/hour, operated, and having a capacity of 30 cubic yard/hour, but operating in this case at only 80% efficiency.

$$300/30 \times .8 = 12.5 \text{ hrs.}$$

Add 2 hours for moving on/off job-site, making a total of 14.5 hours.

$$60 \times 14.5 = 870; \; 870/300 = 2.90/cy$$

Hand trimming of bottoms and sides of trenches is usually required after the machine work. Quantities may be taken either in cubic yards or square feet. The hardness of the soil and the depth of trench are the two main variables affecting the cost.

Example #1:

After machine excavation, hand trim a trench 3 ft deep and 2 1/2 ft wide, 900 ft long in average soil. Assume labor at 15.10/hour and production at 45 square foot/hour. Quantity is based upon the bottom area of the trench only, but includes a nominal amount of side trimming.

$$900 \times 2.5 = 2,250 \text{ sf}$$
$$2,250/45 = 50 \text{ hrs}$$
$$50 \times 15.10 = 755.00$$

A more direct way to do this is:

$$\$15.10 \times 2250/45 \text{ sf} = 755.00.$$

Example #2:

If the above example had been figured by the cubic yard instead of square foot, and assuming that earth to be removed averages 3 1/2" thick and production is 1/2 cubic yard per hour:

$$900 \times 2.5 \times .3/27 = 25 \text{ cy}$$
$$25 \times .5 = 50 \text{ hrs;}$$
$$50 \times 15.10 = 755.00$$

The unit price would be:

$$755/25 = 30.20/cy$$

For convenience and quick reference, the estimator may construct tables similar to the following.

Hand Excavation Labor (At 15.10/hour)		
	Production cy/hr	**Cost/cy Under 5′ deep**
Hardpan	.27	*56.00
Cemented soil	.38	*40.00
Medium clay	.43	*35.00
Dense loam	.48	31.11
Medium loam	.54	28.00
Soft loam	.59	25.45
Crushed rock	.65	23.33
Gravel	.70	21.53
Decomposed granite	.76	20.00
Sand, dry	.92	16.47

Note: Increase unit costs by 25% for over 5′ deep.
*Pneumatic equipment may be necessary

Figure 2.21b

Fine Grading by Hand (At 15.10/hour)				
	Trench Bottoms Production Cost		Slabs on Ground Production Cost	
	sf/hr	/sf	sf/hr	/sf
Cemented soils	36	.42	168	.09
Medium clay	43	.35	196	.077
Medium soil	54	.28	216	.07
Cr rock/gravel	—	—	236	.064
DG/sand	72	.21	270	.056

Figure 2.21c

Backfilling (see Figure 2.12a "h") is accomplished by pushing earth into the unused portions of trenches and compacting it into place. The most important cost-affecting condition is space to work; ample space provides:

1. A convenient stockpile of earth for backfill.
2. Increased use of equipment over hand work.
3. Increased production.

Limited space requires:

1. Moving earth from distant piles.
2. Increased hand work.
3. Low production.

To estimate the cost of backfilling, the estimator may select one of the following conditions as most representative of the project, and use the corresponding unit prices, which are based upon these assumed rates:

Labor	15.10/hr	
Compressor w/tamper	15.00/hr	
Skiploader	40.00/hr	operated
Dump truck	55.00/hr	operated
Machine compactor	35.00/hr	operated
Extra tamper & hose	5.00/hr	

Cost of Backfilling Work Per Cubic Yard			
Condition	Production	Labor	Equip
A—Operated machine compactor plus one man shoveling plus one skiploader ½ time	8 cy/hr	1.89	6.50
B—Moderate working room; 2 men with compressor and 2 tampers plus skiploader ½ time	8 cy/hr	3.78	5.00
C—Mass work as in condition A plus material hauled in by truck ½ time	8 cy/hr	1.89	9.94
D—Close quarters. One man with compressor and tamper plus one man shoveling	4 cy/hr	7.55	3.75
E—Mass work as in condition B plus haul material by truck truck figured ½ time	8 cy/hr	3.78	8.44
F—Handwork as in condition D plus haul material by truck truck figured ¼ time	4 cy/hr	7.55	7.19

Figure 2.21d

Example #1:

In the example (under machine excavation), assume that of the 300 cubic yards of excavation, 125 cubic yards will be used for backfill, and condition D applies.

	Unit Costs		Extensions		
	l	e	Labor	Equip	Total
125 cy	7.55	3.75	944	469	1413

Disposal of excess excavated earth depends, for cost, upon the distance it must be moved. It may:

1. Be used nearby on the construction site, and be pushed there by a dozer, or carried by a loader.
2. Be loaded on trucks, hauled to a distant site and simply dumped.
3. Be loaded on trucks, hauled to a distant site, spread, graded and compacted.

The least expensive method is that whereby the material is simply pushed a short distance (method 1 above) on the site and used for general grading purposes. Sometimes the probable method is not clear and the estimated cost is based upon an average (neither the least nor the most expensive).

If fill is not needed on the jobsite, the cost of disposal may still be slightly reduced by having the grading contractor over-excavate and haul away a quantity equal to the amount which will be left over from footing excavations, see Figure 2.21e.

Figure 2.21e

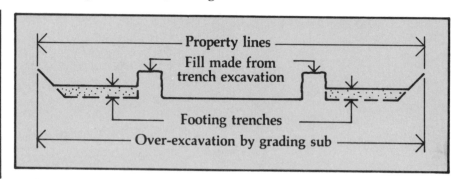

Example #1:

In the previous example (under machine excavation and backfill), 300 cubic yards were excavated and 125 cubic yards were retained for backfill, leaving 175 cubic yards to be disposed of. If the material is pushed aside, spread and compacted on the site, the cost might be calculated as follows:

Skiploader/dozer	65.00/hr	
Compactor	35.00/hr	all operated
Grader (small)	53.00/hr	
	153.00/hr	

Figure 4 hrs. including move on/off time.

$$153.00 \times 4 = 612.00 \text{ total cost}$$

$$\frac{612}{175} = 3.50 \text{ /cy unit cost}$$

Example #2:

If the excess material must be hauled by trucks to a dump site three miles distant, and the cycle time for a truck is thirty-eight minutes, calculate as follows:

$$\frac{175}{10} = 18 \text{ loads}; \quad \frac{18 \times 38}{60} = 11.4 \text{ hrs for one truck}$$

Figure 2 trucks for a total of 6 hrs.

Trucks (10 cy)	2 @ 55 =	110.00/hr
Loader (1 cy)	1 @ 58 =	58.00/hr
		168.00/hr

$$\frac{168 \times 6}{175} = 5.76/cy \text{ unit cost}$$

Note: If part of the loading of trucks can be done by the excavating machine, as it digs the footings, the cost of disposal may be decreased. For instance, if the 58.00/hr loader time can be cut in half, the cost would be:

$$(168 \times 6) - (58 \times 3) = 834, \text{ and } \frac{834}{175} = 4.77/cy$$

Forming is usually figured by the square feet of concrete surface contacted by the form material. For example, in Figure 2.21f the quantity of forms is: (3.0′ + 6.0′) x 2 x 1.5′ = 27 sf.

Sometimes, as in Figure 2.21g, it is more convenient and meaningful to figure a quantity in linear feet. Occasionally it is desirable to figure the number each of forms, grouped by sizes (as column spot footings).

Figure 2.21

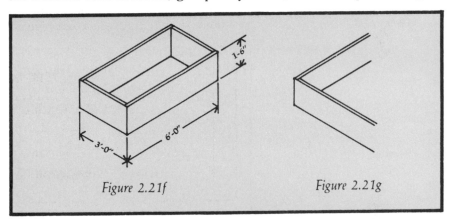

Figure 2.21f   Figure 2.21g

The unit cost includes everything associated with the forming: planning, cutting, panelizing, assembling, erecting, securing, bracing, oiling, removing, cleaning and disposing of materials when the work is completed.

It is a great convenience to the estimator to use unit prices based upon previous calculations and actual field cost records. The following typical unit prices are given as examples to illustrate the systems of estimating presented in this book. It is advisable to regularly recompute and adjust these to keep abreast of labor, material and equipment cost changes (see Section 1.16—Keeping Up with Changes).

For any item of formwork the two most important variables are: the complexity of the form design and the number of times the materials may be reused. The estimator must take into account these two scales.

The rate at which concrete can, or must, be poured sets the pace for form construction. A small quantity of concrete requires one use of forms, but a large quantity of concrete may permit the construction of a portion of the forms, and the reuse of them two, three, four or more times.

Labor tends to decrease with each reuse, because the cutting and panelizing (prefabing), necessary only in the first use, is spread over a larger area; but each time the forms are stripped, moved and re-erected, breakage is likely, so that a small amount of new material must be added and this waste factor increases with each usage.

Example:

Form 24,000 square feet in four sections of 6,000 square feet each.

1st use			6,000 sf
2nd use	add	5% waste	300 sf
3rd use	add	10% waste	600 sf
4th use	add	15% waste	900 sf
		Total	7,800 sf

This is a 30% increase over the original 6,000 sf.

Note: Waste factors differ with different types of forms and forming conditions.

The forming unit prices shown in Figure 2.21h are intended only for use in the examples of typical forming cost estimates which follow.

Formwork for Footings and Foundations

Conditions

Class A - Large quantity of forms, straight or unbroken.
Class B - Medium quantity and medium complexity.
Class C - Small quantity, or broken (many corners, etc.).
Note: These unit prices include any necessary rebar templates.

			Number of Material Uses			
Type 1 - Continuous with pour strip along both sides, per lf			1	2	3	4
		Class				
	Labor	A	2.01	1.94	1.83	1.73
		B	2.22	2.13	2.01	1.91
		C	2.43	2.33	2.20	2.09
	Mat'l		.79	.43	.36	.27
Type 2 - Column (spot) footing. Same detail as above with pour strip 4 sides, per lf of perimeter	**Labor**	A	2.64	2.50	2.38	2.26
		B	2.90	2.76	2.62	2.49
		C	3.19	3.03	2.88	2.74
	Mat'l		1.17	.62	.50	.40
Type 3 - Continuous with one hung screed strip, per lf of footing	**Labor**	A	2.20	2.09	2.12	1.87
		B	2.43	2.33	2.20	2.08
		C	2.69	2.55	2.41	2.04
	Mat'l		1.20	.66	.51	.40

Figure 2.21h

			Number of Material Uses			
			1	**2**	**3**	**4**

Type 4 - Continuous with one staked screed strip, per lf of footing

	Class		1	2	3	4
Labor	A		1.83	1.73	1.65	1.56
	B		2.03	1.93	1.83	1.73
	C		2.22	2.12	2.00	1.90
Mat'l			.40	.22	.16	.14

Type 5 - Continuous with curb, per lf of footing

	Class		1	2	3	4
Labor	A		2.40	2.29	2.15	2.04
	B		2.69	2.55	2.41	2.29
	C		2.95	2.81	2.67	2.52
Mat'l			.91	.51	.38	.30

Type 6 - Continuous with stem wall 6" to 12" deep, per lf of footing

	Class		1	2	3	4
Labor	A		2.56	2.43	2.20	1.87
	B		2.83	2.70	2.56	2.40
	C		3.12	2.95	2.67	2.52
Mat'l			1.38	.77	.57	.44

Type 7 - Continuous **with side forms to 12"** deep, per lf of footing

	Class		1	2	3	4
Labor	A		2.08	1.98	1.87	1.76
	B		2.30	2.18	2.08	1.95
	C		.99	.56	.40	.33
Mat'l			1.10	.70	.60	.50

Type 8 - Continuous with side forms over 12" deep, per sf

		Class	1	2	3	4
14" deep	**Labor**	B	1.90	1.79	1.70	1.60
	Mat'l		.83	.44	.34	.27
16" deep	**Labor**	B	1.94	1.86	1.74	1.65
	Mat'l		.85	.47	.36	.29
18" deep	**Labor**	B	2.00	1.92	1.82	1.73
	Mat'l		.87	.51	.37	.33
20" deep	**Labor**	B	2.08	1.95	1.87	1.76
	Mat'l		.90	.52	.40	.34
22" deep	**Labor**	B	2.11	2.00	1.90	1.79
	Mat'l		.91	.55	.42	.36
24" deep	**Labor**	B	2.18	2.10	1.95	1.87
	Mat'l		.94	.55	.43	.38

Figure 2.21h (cont.)

			Number of Material Uses			
			1	**2**	**3**	**4**
Type 9 - Continuous with curb & side forms, per lf of footing		**Class**				
	Labor	A	2.69	2.55	2.41	2.29
		B	2.95	2.81	2.67	2.51
		C	3.26	3.09	2.94	2.76
	Mat'l		2.30	1.26	.95	.75
Type 10 - Continuous stem wall side forms over 12" deep, per sf						
	Labor	B	3.30	3.13	2.97	2.82
	Mat'l		1.25	.68	.51	.39
Type 11 - Column (spot) footing with staking and screed strip, per lf of perimeter						
	Labor	B	3.00	2.85	2.70	2.55
	Mat'l		.40	.27	.20	.15
Type 12 - Column (spot) footing with sides formed, per sf						
12" deep & less	**Labor**	B	2.10	1.98	1.87	1.76
	Mat'l		.95	.52	.40	.33
14" deep	**Labor**	B	2.16	2.08	1.94	1.86
	Mat'l		.96	.53	.41	.34
16" deep	**Labor**	B	2.21	2.10	1.98	1.87
	Mat'l		.98	.55	.43	.36
18" deep	**Labor**	B	2.34	2.21	2.10	1.98
	Mat'l		.99	.56	.44	.38
20" deep	**Labor**	B	2.40	2.27	2.16	2.03
	Mat'l		1.00	.57	.46	.39
22" deep	**Labor**	B	2.45	2.34	2.21	2.10
	Mat'l		1.01	.59	.47	.40
24" deep	**Labor**	B	2.51	2.40	2.27	2.13
	Mat'l		1.03	.60	.48	.42
Type 13 - Pier forms made up in boxes, per sf						
Up to 10 piers	**Labor**	B	3.38	3.20	3.00	2.80
	Mat'l		2.65	1.61	1.08	.90
10 to 30 piers	**Labor**	B	3.08	2.91	2.76	2.62
	Mat'l		1.61	.88	.65	.52
Over 30 piers	**Labor**	B	2.80	2.65	2.52	2.41
	Mat'l		1.21	.68	.48	.39

Figure 2.21h (cont.)

Slab Edge Forms Cost per Linear Foot

			Number of Material Uses			
			1	2	3	4
2″ deep	Labor	B	1.18	1.13	1.05	1.00
	Mat'l		.20	.13	.08	.07
3″ deep	Labor	B	1.38	1.33	1.24	1.14
	Mat'l		.40	.22	.16	.13
4″ deep	Labor	B	1.57	1.51	1.40	1.34
	Mat'l		.59	.33	.23	.18
6″ deep	Labor	B	1.77	1.70	1.59	1.52
	Mat'l		.79	.43	.33	.26
8″ deep	Labor	B	1.96	1.87	1.77	1.70
	Mat'l		.99	.55	.40	.29
10″ deep	Labor	B	2.16	2.04	1.94	1.83
	Mat'l		1.18	.65	.44	.36
12″ deep	Labor	B	2.35	2.24	2.13	2.01
	Mat'l		1.38	.77	.55	.42

Figure 2.21i

Formwork for Walls (Above Grade)

Conventional wood construction using plywood, studs, wales, and metal ties, cost per sf.

Height			Number of Material Uses			
			1	2	3	4
3'	Labor	B	2.69	2.48	2.28	2.07
	Mat'l		1.16	.73	.60	.57
4'	Labor	B	2.73	2.52	2.31	2.09
	Mat'l		1.20	.75	.62	.59
5'	Labor	B	2.77	2.58	2.35	2.13
	Mat'l		1.22	.77	.64	.61
6'	Labor	B	2.81	2.63	2.40	2.16
	Mat'l		1.24	.79	.67	.63
7'	Labor	B	2.85	2.66	2.45	2.19
	Mat'l		1.27	.83	.70	.65
8'	Labor	B	3.02	2.74	2.52	2.25
	Mat'l		1.35	.88	.75	.67
9'	Labor	B	3.09	2.80	2.57	2.30
	*Mat'l		1.40	.91	.77	.70
10'	Labor	B	3.17	2.87	2.60	2.36
	*Mat'l		1.47	.96	.79	.73
11'	Labor	B	3.21	2.94	2.68	2.40
	*Mat'l		1.50	.98	.81	.75
12'	Labor	B	3.29	3.00	2.71	2.46
	*Mat'l		1.56	1.01	.84	.77
13'	Labor	B	3.47	3.15	2.80	2.55
	*Mat'l		1.60	1.05	.87	.80
14'	Labor	B	3.51	3.20	2.85	2.60
	*Mat'l		1.65	1.08	.90	.84
15'	Labor	B	3.57	3.25	2.92	2.67
	*Mat'l		1.69	1.11	.92	.85
16'	Labor	B	3.60	3.30	3.00	2.72
	*Mat'l		1.75	1.15	.94	.87

*Add erecting and dismantling equipment (crane) for high panels at ground floor. Also, add equipment for *all* forming at second floor and above.

Figure 2.21j

Formwork for Square Columns Per Square Foot

(Conventional plywood & metal clamps)

Uses			Height of Column in Feet			
			8′ to 10′	10′ to 14′	14′ to 16′	16′ to 18′
1	Labor	B	3.43	3.77	4.12	4.46
	Mat'l		2.11	2.18	2.24	2.32
2	Labor	B	3.09	3.43	3.77	4.12
	Mat'l		1.16	1.22	1.30	1.36
3	Labor	B	2.83	3.19	3.52	3.87
	Mat'l		.86	.92	1.00	1.13
4	Labor	B	2.57	2.91	3.26	3.60
	Mat'l		.70	.77	.83	.90
5	Labor	B	2.41	2.74	3.09	3.43
	Mat'l		.62	.68	.76	.82
6	Labor	B	2.24	2.57	2.91	3.26
	Mat'l		.56	.64	.70	.77
7	Labor	B	2.18	2.52	2.77	3.21
	Mat'l		.53	.60	.66	.73
8	Labor	B	2.12	2.47	2.81	3.15
	Mat'l		.50	.56	.64	.70
9	Labor	B	2.09	2.43	2.78	3.12
	Mat'l		.48	.53	.61	.67
10	Labor	B	2.05	2.41	2.74	3.09
	Mat'l		.47	.52	.60	.66

Note: This table is based upon 12″ x 12″ columns; increase or decrease labor (only) by 1% per inch of column size length and width (for 8″ square column, increase labor unit price 4%; for 24″ square column, decrease labor unit price 12%; for 12″ x 14″ column decrease labor unit price 2%).

Figure 2.21k

Formwork for Flat Suspended Slab Soffits

(Plywood on wood joists, beams and shoring 8' high)

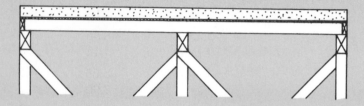

Slab Thickness			Number of Material Uses			
			1	2	3	4
4"	Labor	B	2.29	2.16	2.05	1.96
	Mat'l		1.72	.95	.70	.57
5"	Labor	B	2.40	2.29	2.16	2.05
	Mat'l		1.81	1.00	.74	.60
6"	Labor	B	2.52	2.41	2.29	2.16
	Mat'l		1.89	1.04	.77	.62
7"	Labor	B	2.67	2.52	2.41	2.29
	Mat'l		1.98	1.11	.81	.68
8"	Labor	B	2.81	2.67	2.52	2.41
	Mat'l		2.09	1.14	.84	.70
9"	Labor	B	2.94	2.81	2.67	2.52
	Mat'l		2.20	1.20	.87	.74
10"	Labor	B	3.09	2.94	2.81	2.67
	Mat'l		2.30	1.27	.94	.77
11"	Labor	B	3.25	3.09	2.94	2.81
	Mat'l		2.42	1.33	.98	.81
12"	Labor	B	3.42	3.25	3.09	2.95
	Mat'l		2.55	1.40	1.03	.85

Figure 2.21l

Formwork for Beams Per Square Foot
Beam Bottoms Only, Including Shoring to 8' High

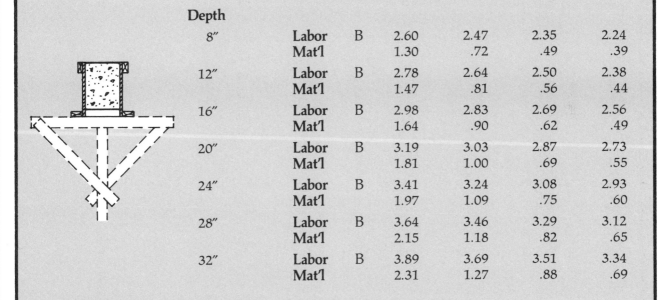

Width			Number of Material Uses			
			1	**2**	**3**	**4**
8"	Labor	B	6.86	6.18	5.55	5.00
	Mat'l		4.14	2.33	1.63	1.27
12"	Labor	B	6.08	5.47	4.93	4.44
	Mat'l		3.90	2.15	1.50	1.17
16"	Labor	B	5.30	4.77	4.29	3.86
	Mat'l		3.56	1.96	1.30	1.07
20"	Labor	B	4.50	4.14	3.64	3.20
	Mat'l		3.22	1.77	1.24	.96
24"	Labor	B	3.74	3.37	3.03	2.73
	Mat'l		2.87	1.59	1.11	.87
28"	Labor	B	3.48	3.13	2.82	2.54
	Mat'l		2.30	1.27	.89	.70

Figure 2.21m

Beam Sides Only Per Square Foot

Depth						
8"	Labor	B	2.60	2.47	2.35	2.24
	Mat'l		1.30	.72	.49	.39
12"	Labor	B	2.78	2.64	2.50	2.38
	Mat'l		1.47	.81	.56	.44
16"	Labor	B	2.98	2.83	2.69	2.56
	Mat'l		1.64	.90	.62	.49
20"	Labor	B	3.19	3.03	2.87	2.73
	Mat'l		1.81	1.00	.69	.55
24"	Labor	B	3.41	3.24	3.08	2.93
	Mat'l		1.97	1.09	.75	.60
28"	Labor	B	3.64	3.46	3.29	3.12
	Mat'l		2.15	1.18	.82	.65
32"	Labor	B	3.89	3.69	3.51	3.34
	Mat'l		2.31	1.27	.88	.69

Figure 2.21n

Figure 2.21

Formwork for Suspended Stairs Per Square Foot

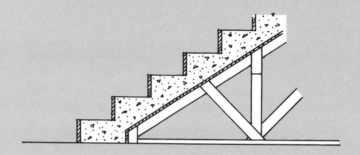

| | | **Number of Material Uses** | | | |
		1	**2**	**3**	**4**
Labor	A	4.75	4.50	4.28	4.00
Mat'l		3.28	1.88	1.41	1.31
Labor	B	6.33	6.00	5.71	5.43
Mat'l		3.64	2.10	1.58	1.46
Labor	C	7.91	7.52	7.14	6.79
Mat'l		4.00	2.30	1.74	1.60

Figure 2.21o

Continuous footing formwork requires of the estimator an assumption of the design that will probably be used by the project-super. Routine questions are:

1. Will the trench sides stand vertically so that side forms are not necessary?
2. Will the top of the footing be above, flush with, or below the finish grade?
3. What is the complexity of the formwork—A, B, or C?
4. How many times may the materials be reused?
5. Based upon answers to the above questions, which form design, Figure 2.21h, most nearly suits the requirements?

The unit prices given in the tables for complexities A, B and C are, themselves, only averages, as indicated in Figure 2.21p, and values may be interpolated between them.

Figure 2.21p

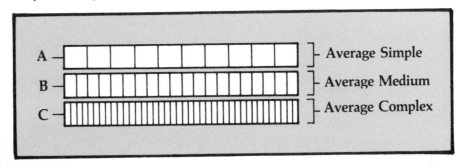

Example:
1900 square feet of Type 8 footing 24" deep, condition B, 3 uses of material - find in table Figure 2.21h:

labor = 1.95 and material = .43

Seldom, however, does an item cost exactly average, so the estimator judges the complexity in this case to be 15% greater than condition B (see Sections 1.8—Judgment and 1.20—Adjusting for Complexity/Simplicity).
The adjusted cost is:

Quantity	Unit Cost		Extensions		
	l	m	L	M	T
1,900 sf	2.24	.49	4,256	931	5,187

Column (spot) footing formwork may be priced out by the linear feet of perimeter, the square feet of contact surface, or by the number each of various sizes (see Figure 2.21h).

Example #1: (by the linear foot)
360 lf of Type 2, condition C, 4 uses.

Q	l	m	L	M	T
360 lf	2.74	.40	986	144	1,130

Example #2: (by the square foot)
540 sf Type 12, condition B, 3 uses, 18" deep.

Q	l	m	L	M	T
540 sf	2.10	.44	1,134	238	1,372

Foundation stem wall formwork over 3' high should be priced out as regular wall forms using Figure 2.21j; those under 3' high are determined on the average by dividing the total square feet by the total linear feet, then dividing again by 2 thus: sf x .5/lf

Example:

5,400 sf and 1,000 lf, condition B, 2 uses. 5,400 x .5/1000= 2.7' high, therefore, use type 10 Figure 2.21h.

Q	l	m	L	M	T
5,400 sf	3.13	.68	16,902	3,672	20,574

Slab edge (header) forms increase in cost per linear foot as the depth increases.

Example:

400 lf, 8" deep, 4 uses, Figure 2.21i.

Q	l	m	L	M	T
400 lf	1.70	.29	680	116	796

Forms for walls increase in unit cost with increased height, because forming material becomes thicker, heavier, more closely spaced and requires more bracing and scaffolding. The cost of detail work, such as forming for openings, bulkheads, chamfer strips and setting inserts are not included in the basic forming cost, but are figured as separate elements.

Example #1:

448 sf, 8' high, 2 uses, ground floor, condition B, but labor is judged at 10% more than the unit prices in Figure 2.21j.

Q	l	m	L	M	T
4,480 sf	3.01	.88	13,485	3,942	17,427

Example #2:

8,960 sf, 16' high, 4 uses, condition B

	Q	l	m	L	M	T
Wall forms	8,960 sf	2.72	.87	24,371	7,795	32,166
Hoisting	8,960 sf		.12		1,075	1,075

Example #3:

800 sf, 10' high, 1 use, condition B (labor judged at 20% and material 10% more than the unit prices in Figure 2.21i).

	Q	l	m	L	M	T
Wall forms	800 sf	3.80	1.62	3,040	1,296	4,336
Hoisting	800 sf		.12		96	96

Column formwork cost varies with cross section dimensions and height.

Example #1:

600 sf, 9' high, 10" square, 5 uses (note that cost is judged to be 2% higher than Figure 2.21k values).

Q	l	m	L	M	T
600 sf	2.46	.63	1,476	378	1,854

Example #2:

1,700 sf, 15' high, 16" square, 8 uses (note that cost is 4% lower than Figure 2.21k values).

Q	l	m	L	M	T
1,700 sf	2.70	.61	4,590	1,037	5,627

Beam formwork is often divided into two parts: (1) bottom forms, including shoring, and (2) side forms. The number of reuses of material, width and depth of beam, and complexity are important variables.

Example:

180 lf, 12" wide and 28" deep, 3 uses, Figure 2.21m and Figure 2.21n

	Q	l	m	L	M	T
Bottoms	180 lf	4.93	1.50	887	270	1,157
Sides	840 sf	3.29	.82	2,764	689	3,453

Plain suspended slab formwork increases in cost with the thickness of slab and height of shoring. Variables for shoring heights will not be considered in the examples given in this book. Figure 2.21i cost values will be used, based on 8' high shoring.

Example #1:
800 sf, 4" slab, 1 use of materials

Q	l	m	L	M	T
800 sf	2.29	1.72	1,832	1,376	3,208

Example #2:
6,000 sf, 7" slab, 2 uses of materials

Q	l	m	L	M	T
6,000 sf	2.52	1.11	15,120	6,660	21,780

Suspended stair formwork may be figured by the gross contact square feet, including soffits, risers and edges, Figure 2.21o.

Condition A - Two or more flights of plain and simple stairs.
Condition B - One flight with one landing, simple to medium complexity.
Condition C - Ornate, with two or more landings, and/or turns.
Note: One flight (one story high), or less should always be figured as condition C.

Example #1:
92 sf (1 flight), 1 use, condition C

Q	l	m	L	M	T
92 sf	7.91	4.00	728	368	1,096

Example #2:
736 sf (8 flights), 4' uses, condition B

Q	l	m	L	M	T
736 sf	5.43	1.46	3,996	1,075	5,071

Example #3:
368 sf (4 flights), 2 uses, condition C

Q	l	m	L	M	T
368	7.52	2.30	2,767	846	3,613

General note about formwork:
The tables and examples given here are of typical items in construction projects. The real challenge to the estimator is to pinpoint the degree of complexity which causes the cost of one item to differ from that of another of the same kind.

In addition to the foregoing typical items of formwork, there will be many others, not so typical such as parapets, spandrel walls, round columns, pile caps, etc. Many of them are closely enough related to the typical items that the estimator may figure their cost by comparison. For instance, pile cap forms are similar in cost to column (spot) footing forms.

Note on complexity/simplicity:
As Figure 2.21p shows, conditions A, B and C are only averages, and the true cost may lie somewhere between, over or under them. The examples indicate that the estimator may select a unit cost on a percentage of variance from one of the three conditions. But percentages are not as clearly visualized as man-days or crew-days. These two examples illustrate the point:

Example #1:

Given a quantity of 1,275 square feet and an average unit price of 3.75:

$$1,275 \text{ sf @ } 3.75 = 4,781$$

Complexity is judged at 18% greater than average:

$$4,781 \times 1.18 = 5,642$$
$$5,642/1,275 = 4.43/\text{sf}$$

Example #2:

Instead of the percentage increase in Example #1, assume that the appropriate cost per crew-hour is 64.00, then 4,781/64 = 75 crew hour, or 9.33 crew days; increase the time to 11 days for complexity, and 11 x 8 x 64 = 5,632; 5,632/1,275 = 4.42/sf

Note on crew hours:

The actual number of workmen of different pay classifications will be determined in the field, yet, the estimator must take into account the *probable* combinations. See Figure 2.4i for convenient crew-hour costs. For a variety of current crew-hour costs, the reader is referred to R.S. Means *Building Construction Cost Data*.

Example #1:

A small quantity, such as 300 square feet, of wall forms could be done by a small crew of, say, 2 carpenters and 1 laborer in 3 days, or 24 crew hours.

$$24 \times 52 = 1,248; \ 1,248/300 = 4.16/\text{sf}$$

Example #2:

A large quantity, such as 6800 square feet, of wall forms could be done by, say, 5 carpenters and 4 laborers in, say, 17 days, or 136 crew hours.

$$136 \times 152 = 20,672; \ 20,672/6,800 = 3.04/\text{sf}$$

Note on forming materials:

Blind use of unit prices for materials can be misleading. Materials used only once or twice may still be used in other parts of the project or have salvage value. In some projects it may be worthwhile for the estimator to analyze, compare and adjust the total estimated cost of form materials in the following manner.

1. If forming materials are figured by the unit cost method, group and summarize them as in Figure 2.21q.

	Uses	Quantity	Unit Price	Total
Footings	3	1,900 sf	.49	931
Stem walls	2	5,400 sf	.68	3,672
Slab edges	4	400 lf	.29	116
Walls	4	8,960 sf	.99	8,870
Beams	3	1,080 sf	.89	961
Columns	8	1,700 sf	.61	1,037
Susp. slabs	2	6,000 sf	1.11	6,660
Stairs	2	368 sf	2.30	846
Miscellaneous	2	1,000 sf	.80	800
TOTAL				23,893

Figure 2.21q

2. Tabulate the form materials as in Figure 2.21r to display the most economical usage and resulting cost. This example is then compared to Figure 2.21q and an adjustment is made to the previous estimate.

Summary of Forming Materials

A Item	B Gross Quant	C Uses	(1) D Net Quant	(2) E Factor	(3) F Lumber BF	G Plywd SF	(4) H Hdwe Factor	I Hdwe $	(5) J Oil Gals
Footings	1,900 sf	3	766	2.25	958	766	.02	38	10
Stem walls	5,400 sf	2	2,970	2.60	4,752	2,970	.06	324	27
Slab edges	400 lf	4	133	1.00	133	—	.02	8	2
Walls	8,960 sf	4	2,981	3.00	5,960	2,981	.06	532	45
Beams	1,080 sf	3	436	3.25	981	436	.05	54	6
Columns	1,700 sf	8	263	2.70	447	263	.12	204	9
Susp. slabs	6,000 sf	2	3,300	3.60	8,580	3,300	.03	180	30
Stairs	368 sf	2	202	7.00	1,212	202	.08	29	2
Miscellaneous	1,000 sf	2	556	3.00	1,100	550	.04	40	5
Totals					24,123	11,468		1,409	136
Unit costs					.60	.50			5.00
					14,473	5,734		1,409	680

Grand total columns F,G,I, and J = 22,296

Explanation of Notes Above Columns

(1) Column B divided by C + 10% waste for each usage of materials
(2) Board feet of lumber required for one sf of forms, see Fig. 2.4g
(3) Columns D x E = F + G
(4) Form hardware unit costs, see Fig. 2.4h; Columns B x H = I
(5) Form oil assumed to cover 200 sf/gal; Column B divided by 200

Comparison of this summary to Figure 2.21p:

Total cost of form materials by unit cost method, Fig. 2.21q	23,893
Total cost of form materials by this summary	22,296
Difference	1,597

Figure 2.21r

Concrete furnishing & placing includes hoisting or pumping, wheeling, dumping, chuting, vibrating, tamping and striking off roughly to finish elevations. Setting screeds, finishing and curing are separate items of work.

The material is considered first because all costs, even labor and equipment, depend upon the characteristics of the material. Transit mix (the rule in most construction projects) constitutes the basis of the following examples. The take-off should identify these typical characteristics of the specified concrete material:

1. Strength of the concrete (2,000 psi, 3,000 psi, etc.)
2. Size of aggregate (max 3/4", 1", 1 1/2", etc.)
3. Water content-slump (3", 4", 5", etc.)
4. Admixtures (air-entrainment, accelerators, retarders, color, etc.)
5. Type of handling equipment required (pumping, for instance, effects the formula and therefore the cost)
6. Probable number of pours (cubic yards per day or half day)

The estimator passes the specific information to area transit-mix companies and obtains firm quotations similar to those in Figure 2.21s. These typical unit prices will be used in the examples given later in this book. Local and state taxes are presumed to be included, as well as specified admixtures.

For current average prices of concrete, the reader is referred to R.S. Means *Building Cost Construction Data*.

Figure 2.21s

Transit Mixed Concrete Material Costs for jobsite/truck				
			Maximum Aggregate Size	
	Pump Mix	¾"	1"	1½"
2,000 psi	63.00	60.90	57.12	54.40
2,500 psi	63.95	60.90	58.00	55.25
3,000 psi	72.93	69.46	66.15	63.00
3,500 psi	78.65	74.90	71.33	67.93

The cost of labor depends upon the volume of concrete to be placed in a day or half day. The number of workmen required varies with the number of cubic yards of concrete and the manner of its placing. Rather than work out the labor cost of each item for the examples given in this book, the following table may be used to find the approximate unit costs for placing (see Figure 2.21t).

Concrete in footings or foundations may be poured directly from transit-mix trucks, or conveyed into the forms by pumping, craning, conveyor belt, or wheelbarrow. A portion of the concrete may be poured directly from trucks, and the remainder by the use of suitable handling equipment.

Example #1:
90 cubic yards of 3000 psi, 1 1/2" max aggregate concrete are to be poured directly from trucks in one 5-hour day. From Figures 2.21s and 2.21t find:

Q	l	m		L	M	T
90 cy	6.45	63.00		584	5,670	6,251

Cost of Pumping Concrete					
Quant CY	Pump Time Hrs	Equip Cost	Equip Cost Per Cy	Labor No. of Men	Cost Per CY
10	4	813	81.30	3	17.42
20	4	"	40.65	3	8.17
30	4	"	27.10	4	7.74
40	4	"	20.33	5	7.26
50	4	"	16.26	6	6.97
60	4	"	13.55	7	6.78
70	4	"	11.61	8	6.45
80	4	"	10.16	8	5.81
90	5	1015	11.28	8	6.45
100	5	"	10.15	8	5.81
110	6	1219	11.08	9	7.13
120	6	"	10.16	9	6.53
130	7	1422	10.94	9	7.04
140	8	1625	11.61	10	8.30
150	8	"	10.83	10	7.74
160	8	"	10.16	11	7.99
170	8	"	9.56	11	7.52
180	8	"	9.03	12	7.74
190	8	"	8.55	12	7.34
200	8	"	8.13	12	6.97
210	12	2438	11.61	14	7.74
220	12	"	11.08	14	7.39
230	12	"	10.60	16	8.08
240	12	"	10.16	16	7.74
250	13	2683	10.73	16	7.43
260	13	"	10.32	16	7.15
270	13	"	9.94	17	7.31
280	14	2844	10.16	17	7.05
290	14	"	9.81	17	6.81
300	14	"	9.48	17	6.58
310	15	3047	9.83	18	6.74
320	15	"	9.52	18	6.53
330	15	"	9.23	18	6.34
340	16	3250	9.56	18	6.15
350	16	"	9.29	18	5.97

One Pump (rows 10–200), Two Pumps (rows 210–350)

Notes: A stand-by pump is included in the above rates. To vary the use of the above table, divide the cost for a given number of hours in the "equip cost" column by the estimated number of cubic yards, thus:

100 cy in 8 hrs would cost $\frac{1625}{100} = 16.25/cy$

Figure 2.21t

Allowance for waste concrete must be added to the quantity of concrete taken off for each item. As a rule, more concrete must be ordered from the mixing plant than the net volume computed from the drawings. Some of the reasons are:

1. Earth under slabs or footings is not graded to perfect elevations.
2. Footing side forms are omitted and concrete is poured against the earth trench sides, which are irregular or over-excavated.
3. Leaking or out of line (bowed) forms.

4. The need to over-order by truckloads to ensure complete pours.
5. The necessity to replace hardened concrete (due to delayed pours) with fresh concrete.

Estimating the quantity of concrete for waste is a matter of judgment and experience. The greatest amount of waste (in terms of percentage) occurs in over-excavated trenches and in small quantity items; the least amount of waste occurs in large quantities of simple and precisely constructed formwork.

The following percentages and rules are offered as guidelines for estimating the amount of waste.

Slabs on ground	6 to 12% see rule (2) below
Footings, formed	4%
Footings poured against earth	10 to 30% see rule (4) below
Foundation walls, formed	11%
Walls, formed	3 to 6% see rule (3) below
Columns	15%
Beams	10%
Suspended slabs	8%
Stairs	15%

Miscellaneous rules:

(1) For small quantities, round out to the next whole cubic yard, thus: if take-off shows 9.07 cy, use 10 cy

(2) For slabs on ground figure approximately ½" extra thickness, thus:

4" slab	12%
6" slab	8%
8" slab	6%

(3) For walls figure approximately ⅜" extra thickness, thus:

6" wall	6%
8" wall	5%
10" wall	4%
12" wall	3%

(4) Use a handy table, such as Fig. 2.21v, to find the percentage of waste concrete in various sizes of over-excavated footings.

Figure 2.21u

Example #2:

120 cy of 2500 psi, 1" aggregate concrete are to be poured in two days (60 cy/day), all pumped; from Figures 2.21s and 2.21t.

	Q	l	m/e	L	M/E	T
Labor	120 cy	6.78	58.00	814	6,960	7,774
Mat'l						
Equip	120 cy		13.55		1,626	1,626

Allowance for Waste Concrete

Where Footings are Cast in Trenches With no Side Forms

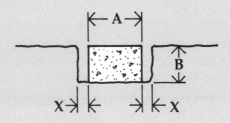

Footing Dimensions in inches		Percentage of increase over net quantity where 2x is:				
A	**B**	**2"**	**3"**	**4"**	**5"**	**6"**
12	8	24	32	42	50	59
12	10	23	31	40	49	57
12	12	22	30	39	48	56
14	10	20	27	35	42	50
14	12	19	26	33	41	49
14	14	18	25	32	40	48
16	12	17	23	30	36	42
16	14	16	22	29	35	41
16	16	15	22	28	34	40
18	12	16	21	27	33	39
18	14	15	20	26	32	38
18	16	15	20	26	32	37
18	18	14	20	25	31	37
20	20	12	18	23	27	33
22	22	12	16	21	25	31
24	24	11	15	19	23	28
30	30	9	12	15	19	22
36	36	7	10	13	16	18

Figure 2.21v

Example #3:

420 cy of 3000 psi, 1″ max aggregate concrete are to be poured in 3 days, 280 cy to be pumped.

Solution:

$\dfrac{420}{3}$ = 140 cy/day; from Figures 2.21s and 2.21t

labor = 8.30/cy; equipment cost at the rate of 140 cy/day = 11.61/cy

	Q	l	m/e	L	M/E	T
Labor ⎱ Mat'l ⎰	420 cy	8.30	69.46	3,486	29,173	32,659
Equip	280 cy		11.61		3,250	3,250

Placing concrete in all other locations, such as slabs on ground, walls, beams, columns and stairs, involves the same considerations as the examples just given. The single greatest variable is the quantity (cubic yards) that may be placed per hour, or half-day, or whole day. The following Figure 2.21w is an example showing typical variations.

Miscellaneous concrete work such as catch basins and manholes, because of their uniformity, may be estimated with the use of tables, such as Figures 2.21x and 2.21y. The costs given in these tables include excavation, backfill, disposal of excess dirt, forming, concrete, setting miscellaneous metal, patching and curing. Reinforcing steel and the furnishing of miscellaneous metal are not included, since they are customarily furnished by subcontractors. Rectangular shapes may be computed by equating them to the nearest square shape in size, thus a 4′ x 6′ basin will cost approximately the same as a 5′ x 5′ basin of the same depth.

Setting screeds for slabs varies in cost with the kind of material composing the sub-base and with the thickness of slab. Figure 2.21aa shows some typical unit costs which will be used in the examples given in this book.

Finishing and curing of slabs are often priced separately, but for convenience they are combined in Figure 2.21bb. The cost of curing varies with the type of curing material used; the examples in Figure 2.21bb are based upon sprayed-on liquid compound.

Pointing and patching of formed surfaces is usually required to fill voids, remove fins and plug form-tie holes. Some conditions affecting the cost are:

1. Whether the areas are concealed or visible.
2. The quality of the forms and the vibrating of the concrete while placing.
3. Accessibility. Note: scaffolding cost is described in Section 2.18— General Conditions.

The range of unit costs (Figure 2.21z) will be used in the examples given in this book.

Rub and grind, or sack finish is a variable quality "architectural" finish of concrete vertical surfaces exposed to view. For the examples in this book, the range of unit costs (Figure 2.21cc) will be used.

Miscellaneous items associated with concrete slabs include sand, gravel, waterproof membrane under slab, expansion joints, keyed construction joints and contraction joints. Costs shown in the tables (Figures 2.21dd, 2.21ee and 2.21ff) will be used in the examples given in this book.

Example Estimate of Typical Structural Concrete Items

Item		Half-Day Production CY	Total Quant CY	l	m/e	L	M/E	T
Slabs on grade	2500 psi-1"	30	120	7.74	58.00	929	6,960	7,889
Walls	3000 psi-1"	40	160	7.26	66.15	1,162	10,584	11,746
Equipment		—	—	—	20.33	—	3,253	3,253
Columns (1)	3500 psi-¾"	11	22	31.67	74.90	697	1,648	2,345
Equipment		—	—	—	73.91	—	1,626	1,626
Beams (1)	3000 psi-1"	38	76	15.28	66.15	1,161	5,027	6,188
Equipment		—	—	—	21.39	—	1,626	1,626
Susp. slabs	3000 psi-¾"	50	250	6.97	69.46	1,743	17,365	19,108
Equipment		—	—	—	16.26	—	4,065	4,065
Stairs (2)	3000 psi-¾"	5	20	17.42	69.46	348	1,389	1,737
Equipment		—	—	—	81.30	—	1,636	1,636
Total						6,040	55,169	61,209

Exceptions to unit prices in Figures 2.21s and 2.21t.

(1) Double the labor on columns and beams because of slow work.

(2) Table value reduced by ½ on the assumption that planned concrete pours may use the equipment to better advantage.

Figure 2.21w

Concrete Manholes (With Concrete Tops)

The following price examples include excavation, backfill, and installing manhole covers, ladder rungs, etc. They do not include reinforcing steel, the furnishing of metal items, fringe benefits and taxes on labor.

Inside Length and Width

Inside Depth		4'x4'	5'x5'	6'x6'	7'x7'	8'x8'	9'x9'	10'x10'
	l	585	765	946	1127	1308	1489	1669
4'	m	172	270	369	468	567	666	767
	e	72	86	100	114	129	144	156
	l	709	918	1127	1336	1546	1755	1966
5'	m	212	328	443	559	675	790	904
	e	81	96	112	127	143	159	177
	l	832	1070	1308	1546	1784	2022	2262
6'	m	252	384	515	646	777	909	1040
	e	90	108	126	144	163	181	198
	l	956	1223	1491	1759	2027	2295	2558
7'	m	293	439	586	733	880	1027	1177
	e	99	118	138	157	177	196	218
	l	1079	1375	1672	1968	2265	2561	2855
8'	m	333	497	660	824	988	1152	1313
	e	108	130	152	174	196	218	239
	l	1203	1528	1853	2178	2503	2827	3151
9'	m	373	553	732	911	1091	1270	1450
	e	117	140	164	187	211	234	260
	l	1316	1680	2033	2386	2740	3094	3450
10'	m	416	611	806	1000	1196	1391	1586
	e	124	150	176	202	228	254	280

Figure 2.21x

Examples of miscellaneous concrete work:

1. Estimate the cost of one manhole having inside dimensions of 6' x 8' x 7' (use 7' x 7' x 7').
2. What would be the cost of two catch basins with inside dimensions of 3' x 3' x 3' and 3' x 3' x 5' (use an average depth of 4').
3. What would be the cost to point and patch 3200 square feet of concrete wall at the rate of 200 square feet per man hour without the use of scaffolding?
4. What would be the cost to rub-and-grind (sack) 800 square feet of concrete wall at the rate of 48 square feet per man hour without the use of scaffolding?
5. Estimate the cost of 10,000 square feet of 4" thick concrete slab on ground consisting of the following elements:
   a. Fine grading the earth subgrade.
   b. 4" of gravel (crushed rock) base course (allowing 20% for shrinkage in volume after compaction).
   c. 6 mil polyethylene film, placed at the rate of 325 square feet per man hour (allowing 10% for overlapping).
   d. 2" of screened sand on top of the film (allowing 20% for compaction).

Figure 2.21y

Concrete Catch Basins (No Tops)

The following price examples include excavation, backfill, and installing of miscellaneous metal. They do not include reinforcing steel, furnishing of miscellaneous metal, fringe benefits or taxes on labor.

Inside Length and Width

Inside Depth			2'x2'	3'x3'	4'x4'	5'x5'
2'	l		143	198	252	306
	m		59	83	107	130
	e		13	17	22	26
3'	l		221	316	411	504
	m		90	131	173	213
	e		25	35	46	56
4'	l		299	434	569	703
	m		121	179	238	296
	e		36	53	70	86
5'	l		377	553	728	902
	m		152	228	303	380
	e		48	70	92	116
6'	l		455	671	887	1101
	m		183	277	371	463
	e		60	88	117	146
7'	l		533	789	1045	1300
	m		215	325	436	546
	e		72	107	142	176
8'	l		611	906	1201	1498
	m		244	372	499	627
	e		83	125	166	208

e. Setting of screeds.
f. 2500 psi, 1" aggregate concrete (allowing 12% for waste).
g. 800 linear feet of keyed construction joint (place 451 linear feet/man-hour).
h. 400 linear feet of ½" x 3-½" expansion joint and sealer.
i. Standard trowel finishing and curing.

See Figure 2.21gg for price-out.

Figure 2.21z

	Hourly Production	Cost of Pointing and Patching Per SF	
		l	m
Maximum	252 sf	.06	.03
Minimum	76 sf	.20	.10

Figure 2.21aa

Cost Per Square Foot
Setting Screeds for Concrete Slabs
Type of Sub Base Material

Slab Thickness	Sand Gravel Or Earth		Concrete		Wood		Sand and Membrane		Steel Deck	
	l	m	l	m	l	m	l	m	l	m
2"	.065	.018	.07	.025	.055	.012	.067	.025	.075	.017
3"	.07	.02	.075	.03	.06	.013	.069	.03	.08	.018
4"	.075	.023	.08	.032	.065	.014	.078	.033	.085	.019
5"	.08	.025	.085	.034	.067	.015	.084	.038	.087	.02
6"	.085	.028	.09	.035	.07	.017	.086	.04	.09	.022
7"	.09	.03	.094	.037	.073	.018	.088	.042	.093	.023
8"	.093	.034	.097	.039	.075	.02	.092	.044	.095	.025
9"	.095	.036	.10	.04	.078	.022	.095	.046	.098	.027
10"	.098	.038	.103	.042	.08	.024	.098	.048	.10	.028
11"	.10	.04	.105	.044	.085	.025	.102	.05	.105	.03
12"	.103	.044	.107	.047	.09	.027	.105	.052	.11	.032
13"	.105	.046	.112	.052	.094	.029	.107	.054	.114	.033
14"	.108	.048	.114	.054	.097	.039	.112	.056	.117	.034
15"	.11	.05	.116	.056	.10	.04	.115	.06	.12	.035

Figure 2.21bb

Cost Per Square Foot
Concrete (Cement) Finishing
Liquid Compound Curing Included

	Daily Production	l	m	t
Float only	660 sf	.21	.021	.231
Broom finish	622 sf	.22	.022	.242
Trowel-standard	560 sf	.25	.025	.275
Trowel-burnished	487 sf	.30	.03	.33
Trowel and abrasive non-slip	415 sf	.33	.13	.46
Trowel and salt finish	340 sf	.41	.09	.50
Trowel and exposed aggregate	280 sf	.49	.16	.65

Figure 2.21cc

Cost to Rub and Grind or Sack Per Square Foot

	Hourly Production	l	m
Maximum	69 sf	.25	.08
Minimum	27 sf	.65	.16

Labor and Equipment Cost
To Place Aggregate for Base Course Under Slabs

	Large Areas And Under Paving				Under Floor Slabs			
	Per cy		Per sf		Per cy		Per sf	
	l	e	l	e	l	e	l	e
2″	2.50	8.00	.016	.051	10.37	11.80	.065	.07
3″	2.24	7.15	.021	.065	8.88	11.25	.082	.10
4″	1.86	6.19	.023	.078	7.37	8.58	.092	.108
6″	1.48	4.95	.026	.092	5.92	7.49	.109	.139
8″	1.37	4.43	.034	.111	5.43	6.80	.135	.169
10″	1.22	4.00	.038	.124	5.08	6.08	.150	.187
12″	1.12	3.72	.042	.138	4.43	5.62	.164	.21

Figure 2.21dd

Typical Material Costs

		Per cy	Per sf	Per lf
Fill sand (unscreened)		10.19		
Fill sand (screened)		11.65		
Class II base		13.00		
Fill dirt		8.74		
Decomposed granite		10.19		
Pea Gravel		20.38		
¾″ to ½″ crushed rock		17.47		
Topsoil		10.19		
Polyethylene film	4 mil		.026	
	6 mil		.033	

Figure 2.21ee

Cost Per Linear Foot
Expansion Joints and Sealer in Slabs

	Exp. Jts. No Sealer		Sealer Only		Exp. Jts. and Sealer	
	l	m	l	m	l	m
⅜″ x 3½″	.26	.143	.13	.078	.39	.22
⅜″ x 4½″	.273	.176	.13	.078	.403	.254
⅜″ x 5½″	.286	.215	.13	.078	.416	.294
½″ x 3½″	.312	.195	.156	.104	.468	.299
½″ x 4½″	.325	.234	.156	.104	.481	.338
½″ x 5½″	.338	.273	.156	.104	.494	.377
½″ x 6½″	.35	.325	.156	.104	.506	.429
½″ x 7½″	.364	.39	.156	.104	.520	.494
½″ x 8½″	.377	.455	.156	.104	.533	.559
½″ x 9½″	.39	.52	.156	.104	.546	.624
½″ x 10½″	.403	.585	.156	.104	.559	.689
½″ x 11½″	.416	.65	.156	.104	.572	.754
¾″ x 4½″	.455	.325	.234	.156	.689	.481
¾″ x 5½″	.468	.39	.234	.156	.702	.546
¾″ x 7½″	.494	.52	.234	.156	.728	.676
¾″ x 9½″	.52	.624	.234	.156	.754	.780
¾″ x 11½″	.546	.715	.234	.156	.780	.871

Figure 2.21ff

Example of Miscellaneous Concrete Work Price-Out

	Quantity		l	m/e	L	M/E	T	REF.
1. Manhole 6' x 8' x 7'	1	ea	1,759.00	890.00	1,759	890	2,649	2.21x
2. Catch basins 3' x 3' x 4'	2	ea	434.00	232.00	868	464	1,332	2.21y
3. Point and patch	3,200	sf	.075	.037	240	118	358	2.21z
	200 sf/hr							
4. Rub and grind	800	sf	.36	.045	288	36	324	2.21cc
	40 sf/hr							
5. Slab on ground								
Fine grade, medium soil	10,000	sf	.07	—	700	—	700	2.21c
4" thick crushed rock base course	152	cy	7.37	26.05	1,120	3,960	5,080	2.21dd/ee
Poly film 6 mil	11,000	sf	.053	.033	583	363	946	2.21ee
	325 sf/hr							
2" screened sand on poly film	76	cy	10.37	23.45	788	1,782	2,570	2.21dd/ee
Set screeds 2" sand & memb	10,000	sf	.067	.025	670	250	920	2.21aa
Concrete 2500 psi-1"	141	cy	8.30	58.00	1,170	8,178	9,348	2.21s/t
	70 cy/day							
Keyed construction joints	800	lf	.40	.65	320	520	840	*
	45 lf/hr							
Expansion joints and sealer	400	lf	.468	.299	187	120	307	2.21ff
Trowel and cure, standard	10,000	sf	.25	.025	2,500	250	2,750	2.21bb

*Obtained quotation from material supplier

Figure 2.21gg

2.22
Precast concrete

Precast concrete is concrete which is cast in some location other than its final position. After curing, a precast concrete slab, column, beam, or whatever, is picked up, moved, placed and erected in its final position and secured.

The designer chooses the precast method for one or a combination of the following reasons:

1. Economy in cost.
2. Economy in construction time.
3. Economy in working space.
4. Architectural effect.
5. Where casting-in-place is impractical.

Estimating of precast concrete work is a matter of visualizing and pricing out each step in the casting and erecting process.
Typical items and elements are:

1. Casting
   a. Engineering (for stresses and pick-up point locations).
   b. Temporary casting slab (or use building floor slab).
   c. Forming of edges.
   d. Bond breaker on casting slab.
   e. Embedded items , lifting and bracing inserts, welding inserts, bolts, doorframes, etc.
   f. Reinforcing steel (by subcontractor).
   g. Concrete, placing and vibrating.
   h. Concrete finishing and curing.

2. Hauling to jobsite (if cast off-site).
   a. Loading on trucks.
   b. Hauling.
   c. Unloading and stockpiling at jobsite.
3. Erecting in place.
   a. Rigging (riggers & foremen).
   b. Lifting and moving (crane & operator).
   c. Bracing.
   d. Leveling, shimming and grouting underneath.
4. Securing and finishing.
   a. Welding or bolting.
   b. Grouting between panels.
   c. Pointing up and touching up the finish surfaces.

It is a good idea to price out separately the four items: casting, hauling (if any), erecting and securing, in order to compare to sub-bids which may include one or a combination of these operations.

The following pertains to precast (tilt-up) wall panels but, with variations, may be applied to precast columns, beams, balcony railings, facia panels, floor panels, etc. Figure 2.22a is an example of a complete estimate and it is followed by an explanation of each item.

1. *Engineering* cost depends upon the number of different sizes, shapes and thicknesses of panels. In this example, if engineering time is estimated at 32 hrs and 52.00/hr, then:

$$\frac{32 \times 52}{128} = 13.00/panel$$

Since this is not subject to payroll taxes, it is listed under the M/E column.

2. *Casting slab* is required in this example, because the building floor slab is not available. Similar size panels may be "pancaked", and in this

Example Precast (Tilt-Up) Concrete Wall Panel Estimate

Casting	Quant	l	m/e	L	M/E	T	Ref
1. Engineering	128 ea	—	13.00	—	1,664	1,664	Sec. 2.21
2. Casting slab	5,000 sf	.71	.88	3,550	4,400	7,950	Fig. 2.21i
3. Edge forms	5,700 lf	1.82	.31	10,374	1,767	12,141	
4. Bondbreaker	16,000 sf	.029	.03	464	480	944	
5. Setting weld plates	552 ea	6.04	—	3,334	—	3,334	
6. Setting lift inserts	512 ea	4.53	4.50	2,319	2,304	4,623	
7. Setting brace inserts	512 ea	3.63	4.50	1,859	2,304	4,163	
8. Setting miscellaneous bolts	384 ea	3.02	—	1,160	—	1,160	
9. Concrete, 3000 psi-1"	306 cy	5.81	66.15	1,778	20,242	22,020	Fig. 2.21s/t
10. Finish and cure	16,000 sf	.25	.025	4,000	400	4,400	Fig. 2.21bb
			Subtotal	28,838	31,897	60,735	
Erecting							
11. Crane, operated-45 ton	51 hrs	—	120.00	—	6,120	6,120	Fig. 2.10a
12. Riggers and foreman	52 hrs	—	110.00	—	5,720	5,720	
13. Bracing and leveling	256 hrs	7.37	11.70	1,887	2,995	4,882	
14. Shimming and grouting	128 ea	13.00	2.50	1,664	320	1,984	
			Subtotal	3,551	15,155	18,706	
Securing							
15. Welding & weld plates	256 ea	4.75	1.60	1,216	410	1,626	
16. Grout vertical joints	1,638 lf	2.88	.255	4,717	418	5,135	
17. Point and patch (430 sf/hr)	32,000 sf	.035	.003	1,120	96	1,216	
18. Scaffolding (rolling)	4 mos	110.00	73.00	440	292	732	
			Subtotal	7,493	1,216	8,709	
			Grand Total	39,882	48,268	88,150	

Figure 2.22a

example there are thirty-two basic sizes, so the quantity of casting slab may be computed this way:

$$\frac{16,000}{128} \times 32 = 4,000 \text{ sf - the minimum area for casting}$$

But in order to have working room around the panels, increase the area to 5,000 square feet. Construct a 3″ thick concrete slab-on-ground using transit-mix costing 45.00 per cubic yard. Figure to demolish and dispose of the slab when precast concrete work is completed, using a breaker-loader and two dump trucks of ten cubic yard capacity. Figure the cost of slab and its disposal as follows:

	Quant	l	m/e	L	M/E	T	Ref
Fine grade	5,000 sf	.07		350		350	2.21c
Form edges (1 use)	900 lf	1.38	.40	1,242	360	1,602	2.21i
Set screeds	5,000 sf	.07	.02	350	100	450	2.21aa
Concrete	52 cy	6.78	45.00	353	2,340	2,693	2.21s
Finish & cure	5,000 sf	.25	.025	1,250	125	1,375	2.21bb
Demo & remove	8 hr		185.00		1,480	1,480	2.10a
Totals				3,545	4,405	7,950	
Unit costs per sf				.71	.88		

3. *Edge forms* for panels may be priced out as for slabs on ground.

   In this example $\frac{128}{32} = 4$ uses (see Figure 2.21i 6″ thick).

   However, increase unit costs 20% for extra work due to keys, protruding rebar, etc. Confirming the labor by man-hours, (see Section 2.7—Man-Hours and Pricing Out), the average time to construct edge forms in this example, where carpenter pay is 18.13/hr, is 5700 x 1.82/128÷18.13 = 4.47 hrs. That is to say, one man will form and strip one average panel in approximately 4 1/2 hours (or two men in 2 1/4 hours). This production seems realistic.

4. *Bond breaker* is a compound which prevents the precast concrete panels from bonding to the casting slab. In this example the cost is 4.55/gal, the coverage is 150 sf/gal, and the labor production is 600 sf/hr.

5. *Setting weld plates*, which may involve attaching to forms or rebar, averages in this example 3/hr. No material cost is included because they will be supplied by the miscellaneous metal subcontractor.

6. *Setting lift inserts*, because of the required accuracy, may take fifteen minutes each. The material being of stock manufacture, is included in this example at an assumed unit cost of 4.50 each.

7. *Setting brace inserts* may be slightly less in cost than lift inserts because strength and accuracy are not quite as important. Here, twelve minutes are allowed for each one, and 18.13 x 12/60 = 3.62 each.

8. *Setting miscellaneous bolts* for ledger angles and other attachments requires almost the same care as setting brace inserts. Ten minutes each is allowed. Bolts are to be furnished by subcontractors.

9. *Concrete purchasing and placing* may be estimated by the use of Tables 2.21s and 2.21t;

   figure $\frac{128}{32} = 4$ pours of 76.5 cy each

10. *Finishing and curing* may be estimated by the use of Figure 2.21bb, standard trowel finish.

11. *Erecting panels by crane* usually proceeds at the rate of one to four panels per hour. In this example three panels per hour is the assumed production. The average panel weighs about five tons; experience indicates the use of a forty-five ton crane, and the current operated rental rate, per Figure 2.10a is 120.00/hr:

$$\frac{128}{3} = 42\text{-}2/3 \text{ hours;}$$

rounding out to 43 hours and adding 8 hours for moves on/off the jobsite, the total estimated time is 51 hours.

12. *Riggers and foreman* must be paid to the nearest half-day of work; the nearest half-day beyond the crane rental time of fifty-one hours is fifty-two hours. Riggers are not usually carried on the contractor's payroll, but are "rented" with the equipment at gross pay level, thus the cost is carried in the M/E column. In this example the following pay rates are used:

riggers	2 @ 35.00	=	70
rigger foreman	1 @ 40.00	=	40
total			110/hr

13. *Bracing and leveling* requires a crew working continuously as the erecting work proceeds. In this example two carpenters are figured:

$$\frac{2 \times 18.13 \times 52}{256} = 7.37/\text{each brace}$$

Allow 11.70/month each for brace rental.

14. *Shimming & grouting* under the panels proceeds with the panel erection work. Two men at a crew-hour of 32.00 are figured for fifty-two hours; also a small amount of material is provided.

Labor is: $\frac{52 \times 32}{128} = 13/\text{panel.}$

15. *Welding*, in this example, requires 1/4 hour at each connection; labor rate is 19.00/hr; material and equipment are as shown. Scaffolding is a separate cost item.

16. *Grouting of vertical joints* 2" thick with cement grout is figured at six linear feet per hour and 3.00 per cubic foot, thus:

$$\text{labor } \frac{17.29}{6} = 2.88/\text{lf}$$

$$\text{mat'l } .5 \times .17 \times 3.00 = .255/\text{lf}$$

17. *Pointing and patching* both faces of the panels after welding and removal of braces, to plug up holes and dress up imperfections, is figured in this example at the rate of 1/4 hour per panel side:

$$\frac{128 \times .25 \times 2 \times 17.29}{32,000} = .035/\text{sf}$$

A small amount of material is also included.

18. *Scaffolding* for welding, grouting, pointing and patching, in this example, consists of rolling scaffolding units. Monthly rental of scaffolding towers may be figured this way:

welding	1 tower-month
grouting	2 tower-months
pointing	1 tower-month
Total	4 tower-months

2.23 Rough carpentry

Rough carpentry is generally structural and not exposed to view. However, there are items intermediate between rough and finish carpentry, such as siding and non-milled trim. "Millwork" is only an approximate dividing line.

The take-off begins with those elements which are *structural*, such as beams, girders and columns; next, *supportive*, such as framing, studding, joists, rafters; finally, *covering*, such as sheathing, siding, boarding and trim.

The order of taking off is not necessarily that in which the structure will actually be built (from the ground up). The order is the estimator's preference. In these examples, the rough carpentry take-off proceeds *from the roof downward*.

Typical categories and items include:

1. Framing
    A. Roof framing
        a. Facia
        b. Ledgers
        c. Plates
        d. Blocking
        e. Joists
        f. Rafters
        g. Purlins
    B. Floor framing
        a. Ledgers
        b. Girders
        c. Beams
        d. Joists
        e. Blocking
    C. Ceiling framing
        a. Ledgers
        b. Joists
        c. Blocking
    D. Wall framing
        a. Sills
        b. Blocking
        c. Plates
        d. Studding
        e. Posts
    E. Miscellaneous framing
        a. Blocking
        b. Nailers
        c. Grounds
        d. Stripping
    F. Stair framing & blocking
2. Covering
    A. Plywood
        a. Roof sheathing
        b. Wall sheathing
        c. Shear panels
        d. Sub flooring
    B. Siding
        a. Exterior walls
        b. Interior walls
3. Trim
        a. Exterior
        b. Interior

4. Miscellaneous
    a. Wood fence
    b. Trellis
    c. Alterations
5. Rough hardware
    a. Nails
    b. Bolts
    c. Framing connectors

Computing Board Feet of Lumber
Board Feet Per Linear Foot
of Standard Nominal Sizes

Nominal Size	bf/lf	Nominal Size	bf/lf	Nominal Size	bf/lf	Nominal Size	bf/lf
1x2	.17	2x2	.34	3x3	.75	4x4	1.34
1x3	.25	2x4	.67	3x4	1.00	4x6	2.00
1x4	.34	2x6	1.00	3x6	1.50	4x8	2.67
1x6	.50	2x8	1.34	3x8	2.00	4x10	3.34
1x8	.67	2x10	1.67	3x10	2.50	4x12	4.00
1x10	.83	2x12	2.00	3x12	3.00	4x14	4.67
1x12	1.00	2x14	2.34	3x14	3.50	4x16	5.34
6x6	3.00	8x8	5.34	10x10	8.34	12x12	12.00
6x8	4.00	8x10	6.67	10x12	10.00	12x14	14.00
6x10	5.00	8x12	8.00	10x14	11.67	12x16	16.00
6x12	6.00	8x14	9.34	10x16	13.34	12x18	18.00

Figure 2.23a

The following tables, Figure 2.23b, of labor unit costs will be used in the examples in this book.

Rough Carpentry Labor Unit Costs
Per 1000 Board Feet

Description	Dimensions	Simple A	Medium B	Complex C
Wall sills, bolted	2x3 , 2x4	455	500	550
	2x6 , 2x8	416	458	503
Wall sills, shot	2x3 , 2x4	416	458	503
	2x6 , 2x8	390	429	472
Wall sills, blocking	2x3 , 2x4	455	500	550
Wall blocking	2x6 , 2x8	429	472	519
Wall diagonal brace	1x4 , 1x6	975	1073	1180
	2x3 , 2x4	728	800	881
	2x6 , 2x8	702	772	850
Wall studding	2x3 , 2x4	325	358	394
	2x6 , 2x8	293	322	355
	3x3 , 3x4	293	322	355
Wall top plates	2x3 , 2x4	390	429	472
	2x6 , 2x8	358	394	432
	3x3 , 3x4	358	394	432
Wall posts	3x3 , 3x4	481	529	582
	3x6 , 3x8	455	500	551
	4x4 , 4x6	416	458	503
	6x6 , 6x8	364	400	441
	8x8 , 8x10	338	372	410
	10x10 , 10x12	325	358	394

Figure 2.23b

Rough Carpentry Labor Unit Costs Per 1000 Board Feet				
Description	Dimensions	Simple A	Medium B	Complex C
Wall headers	4x4 , 4x6	559	615	676
	4x8 , 4x10	468	515	567
	4x12 , 4x14	403	443	488
	6x10 , 6x12	358	394	433
Dock bumpers, bolted	2x6 , 2x8	442	486	534
	3x6 , 3x8	429	472	519
	4x4 , 4x6	403	445	488
	6x6 , 6x8	351	386	425
Door bucks	2x6 , 2x8	442	486	534
	3x6 , 3x8	429	472	519
	4x4 , 4x6	403	443	488
	6x6 , 6x8	360	390	430
Underpinning posts	2x3 , 2x4	797	876	963
	4x4 , 4x6	728	800	881
	6x6 , 6x8	650	715	787
	8x8 , 8x10	585	644	709
Underpinning stud & brace	2x3 , 2x4	533	599	645
	2x6 , 2x8	481	529	582
	3x3 , 3x4	455	500	551
	3x6 , 3x8	416	457	503
Beams, 1st & 2nd floor	3x10 , 3x12	403	443	488
	3x14 , 3x16	390	429	472
	4x4 , 4x6	455	500	551
	4x8 , 4x10	403	443	488
	4x12 , 4x14	390	429	472
	6x6 , 6x8	403	443	488
	6x10 , 6x12	364	400	441
	6x14 , 6x16	350	362	425
	8x12 , 8x14	338	372	410
	10x10 , 10x12	325	358	394
	10x14 , 10x16	319	350	386
	12x12 , 12x14	293	322	356
Beams above 2nd floor	3x10 , 3x12	416	458	503
	3x14 , 3x16	403	443	488
	4x4 , 4x6	468	515	567
	4x8 , 4x10	416	458	503
	4x12 , 4x14	403	443	488
	6x6 , 6x8	416	458	503
	6x10 , 6x12	390	429	472
	6x14 , 6x16	364	400	441
	8x8 , 8x10	364	400	441
	8x12 , 8x14	358	394	432
	10x10 , 10x12	350	386	425
	10x14 , 10x16	338	372	410
	12x12 , 12x14	325	358	394
Floor ledgers, bolted	3x6 , 3x8	559	615	676
	3x10 , 3x12	533	586	645
	3x14 , 3x16	480	529	583
Floor blocking	2x3 , 2x4	533	586	645
	2x6 , 2x8	494	543	598
	2x10 , 2x12	480	529	582
	2x14 , 2x16	468	515	567
	3x3 , 3x4	494	543	598
	3x6 , 3x8	480	529	582
	3x10 , 3x12	468	515	567
	3x14 , 3x16	455	500	550
	4x4 , 4x6	494	543	598
	4x8 , 4x10	468	515	567

Figure 2.23b (cont.)

Rough Carpentry Labor Unit Costs
Per 1000 Board Feet

Description	Dimensions	Simple A	Medium B	Complex C
Floor blocking (cont.)	4x12 , 4x14	455	500	550
	6x6 , 6x8	429	472	519
	6x10 , 6x12	416	458	503
	6x14 , 6x16	403	443	488
Floor X bridging	2x3 , 2x4	975	1073	1180
Floor joists	2x6 , 2x8	358	394	433
	2x10 , 2x12	325	358	394
	2x14 , 2x16	299	329	362
	3x6 , 3x8	325	358	394
	3x10 , 3x12	300	329	361
	3x13 , 3x16	293	322	355
	4x8 , 4x10	316	347	382
	4x12 , 4x14	293	322	355
Roof ledgers, bolted	3x6 , 3x8	585	644	709
	3x10 , 3x12	559	615	676
	3x14 , 3x16	468	515	567
Roof blocking	2x3 , 2x4	546	600	660
	2x6 , 2x8	520	572	629
	2x10 , 2x12	494	543	598
	2x14 , 2x16	480	529	582
	3x3 , 3x4	520	572	629
	3x6 , 3x8	494	543	598
	3x10 , 3x12	416	529	582
	3x14 , 3x16	468	515	567
Roof X bridging	2x3 , 2x4	1024	1127	1240
Roof joists	2x6 , 2x8	377	415	456
	2x10 , 2x12	338	372	410
	2x14 , 2x16	325	358	394
	3x6 , 3x8	338	372	410
	3x10 , 3x12	325	358	394
	3x14 , 3x16	300	329	362
	4x8 , 4x10	325	358	394
	4x12 , 4x14	300	329	362
Roof cant strips	3x3 , 3x4	1131	1244	1367
Ceiling joists	2x3 , 2x4	390	429	472
	2x6 , 2x8	358	394	433
	2x10 , 2x12	325	358	394
Ceiling blocking	2x3 , 2x4	481	529	582
	2x6 , 2x8	455	500	551
	2x10 , 2x12	442	486	534
Ceiling hangers	2x3 , 2x4	533	586	645
	2x6 , 2x8	494	543	598
Stair stringers	2x10 , 2x12	975	1072	1180
	2x14 , 2x16	897	987	1084
	3x10 , 3x12	806	887	975
	3x14 , 3x16	767	844	928
Stair blocking	2x10 , 2x12	728	800	881
	2x14 , 2x16	650	715	787
	3x10 , 3x12	650	715	787
	3x14 , 3x16	598	658	724
Decking, T&G	2x3 , 2x4	325	358	394
	2x6 , 2x8	293	322	354
	3x3 , 3x4	293	322	354
	3x6 , 3x8	277	304	334
	1x4 , 1x6	390	429	472
Grounds	1x2 , 1x3	1512	1662	1829

Figure 2.23b (cont.)

Rough Carpentry Labor Unit Costs Per 1000 Board Feet				
Description	Dimensions	Simple A	Medium B	Complex C
Stripping	1x2 , 1x3	975	1073	1180
	1x4 , 1x6	728	800	881
Siding, bd & batt	1x8 , 1x10	390	429	472
	1x12	228	394	433
Nailers on roof deck	2x4 , 6x6	1200	1320	1452
Roof sheathing, plywood	½"	176	293	321
	⅝"	281	312	343
Wall sheathing, plywood	⅜"	281	312	343
	½"	299	332	364
	⅝"	316	351	386
Shear wall, plywood	⅜"	299	332	364
	½"	316	351	386
	⅝"	333	371	408
Floor sheathing, plywood	½"	176	293	321
	⅝"	281	312	343
	¾"	299	332	364

Figure 2.23b (cont.)

It is desirable that the take-off be in the form of a complete and accurate lumber list suitable for sending out to suppliers for quotations. The various species and grades of wood, sizes and number of pieces, surfacing and board feet must be shown. Figure 2.23c is a brief example. When the quantity of rough hardware is large the list may be sent to suppliers for quotations. Figure 2.23d is an example of a hardware list. The labor to install the hardware is included in the unit cost to install the lumber, Figure 2.23b.

Actual lumber and hardware lists may be several pages in length. The estimator will be interested in the *total* price, including tax and delivery to the jobsite. He will then, using the same lumber list, make groupings for pricing out the labor. The following example of price-out, Figure 2.23e, incorporates both the Figure 2.23c lumber list and the Figure 2.23d rough hardware list. It uses the unit prices from Column B in Figure 2.23b, which include foremen, layout work, cutting, erecting and securing in place.

Labor is not figured for the waste material shown on the lumber list. Waste material is provided to compensate for mistakes in cutting, theft, warpage etc.

The amount allowed for waste depends upon the type of material, the clarity of the drawings and the exactitude of the take-off (a very tight take-off may deserve a larger waste factor than a loose one).

Notice that in the price-out, unit costs are converted from Mbf to bf; example:

$$500/Mbf = .50/bf$$

Explanation of the price-out example Figure 2.23e:
The total material cost of lumber, plywood and rough hardware, as quoted by suppliers includes tax and delivery to the jobsite. The quantities are grouped for the purpose of estimating the labor cost. Beams, for instance, are put together in one group, but in jobs where they vary considerably in dimensions and length, they may need to be priced separately. Even so, most elements in a group vary slightly in

Description				Size	No Pcs	Length	Bd Ft
Example of Lumber List							
Roof fascia	const	grd	SA	3x10	Ran	400	1,000
Roof ledgers	const	"		3x10	Ran	320	800
Roof beams	struc	"		6x12	4	20	480
Roof beams	struc	"		4x10	6	16	320
Roof joists	const	"		2x10	100	20	3,340
Roof joists	const	"		2x10	250	16	6,680
Roof blocking	std	"		2x10	Ran	600	1,000
Roof X bridging	std	"		2x3	Ran	1,500	750
Floor ledgers	const	"		3x14	Ran	400	1,400
Floor beams	struc	"		6x14	4	20	560
Floor beams	struc	"		8x8	6	16	512
Floor joists	const	"		2x14	100	20	4,667
Floor joists	const	"		2x14	250	16	9,333
Floor blocking	std	"		2x14	Ran	600	1,400
Floor X bridging	std	"		2x3	Ran	1,500	750
Ceiling ledgers	const	"		2x6	Ran	800	800
Ceiling joists	const	"		2x6	350	20	7,000
Ceiling blocking	std	"		2x6	Ran	1,600	1,600
Wall sills	const	"	PT	2x4	Ran	2,000	1,333
Wall studding	const	"		2x4	2000	8	10,666
Wall blocking	std	"		2x4	Ran	2,000	1,333
Wall top plates	const	"		2x4	Ran	4,000	2,666
Wall bracing	const	"		1x4	24	14	112
Miscellaneous stripping	std	"		1x3	Ran	1,600	400
Miscellaneous grounds	std	"		1x2	Ran	1,800	200
Miscellaneous waste	std	"		2x4	Ran	2,100	1,400
Total							60,502 bf
Plywood							
roof sheathing				⅝"	4' x 8'	320	10,240
wall sheathing				½"	4' x 8'	80	2,560
shear panels				½"	4' x 8'	60	1,920
floor sheathing				⅝"	4' x 8'	320	10,240
Total							24,960 sf

Notes: All lumber is Douglas Fir s4s. All plywood is
Douglas Fir CDX Struc. I. PT = pressure treated.
SA = select for appearance. RAN = random length.

Figure 2.23c

cost so that a unit price is actually an average; for example, Figure 2.23b shows different unit costs for 2 x 6, 2 x 10 and 2 x 14 roof joists. Taken separately, the following quantities, under Column A, would cost:

2 x 6	3,600 bf @	.377 =	1,357
2 x 10	2,700 bf @	.338 =	913
2 x 14	4,200 bf @	.325 =	1,365
Total	10,500 bf		3,635

Each of the unit costs is an average, to a more or less fine degree; yet, the estimator might group all roof joists together and use an average unit cost of coarser degree, as suggested in the totals above, thus:

2 x 6, 2 x 10 & 2 x 14   10,500 bf @ .346 = 3,633

The degree of fineness or courseness in the unit costs depends upon the estimator's judgment, as well as the degree of simplicity or complexity (Columns A, B, C) selected.

Equipment, such as fork lifts, may be needed for hoisting lumber and plywood to upper floors and roof. In this example, 70,000 board feet (including plywood) will be hoisted to and above the second floor level. This will probably be done at different time intervals and it is estimated that forty-eight hours will be required at the rate of 70.00/hr, thus:

$$\frac{48 \times 70}{70,000} = .048/bf$$

The total labor cost is at base pay scale level and does not include payroll taxes and fringe benefits. These will be considered in Section 2.27—Fringe Benefits/Payroll Taxes.

Figure 2.23d

Example of Rough Hardware List		
Description	**Size**	**Quantity**
Joist hangers	2 x 10	350 ea
Joist hangers	2 x 14	350 ea
Tie straps	¼ x 2 x 30	28 ea
Post caps	4 x 4	18 ea
Post bases	4 x 4	18 ea
Anchor bolts	⅝ x 12	120 ea
Anchor bolts	½ x 10	80 ea
Machine bolts	½ x 8	60 ea
Machine bolts	⅝ x 10	100 ea
Lag bolts	⅜ x 6	75 ea
Square washers	½	180 ea
Square washers	⅝	220 ea
Nails, box	8d	600 lbs
Nails, box	16d	300 lbs

Example of Rough Carpentry Price-Out

Description		Quantity		l	m	L	M	T
All lumber per list & quote		60,502	bf	—	ls	—	29,948	29,948
All plywood per list & quote		24,960	sf	—	ls	—	10,982	10,982
All hardware per list & quote			ls	—	ls	—	2,688	2,688
Labor-roof facia	3x10	1,000	bf	.50	—	500	—	500
Labor-roof ledgers	3x10	800	bf	.615	—	492	—	492
Labor-roof beams	6x12 4x10	800	bf	.444	—	355	—	355
Labor-roof joists	2x10	10,020	bf	.372	—	3,727	—	3,727
Labor-roof blocking	2x10	1,000	bf	.543	—	543	—	543
Labor-roof X bridging	2x3	750	bf	1.127	—	845	—	845
Labor-floor ledgers	3x14	1,400	bf	.529	—	741	—	741
Labor-floor beams	6x14 8x8	1,072	bf	.362	—	388	—	388
Labor-floor joists	2x14	14,000	bf	.329	—	352	—	352
Labor-floor blocking	2x14	1,400	bf	.529	—	741	—	741
Labor-floor X bridging	2x3	750	bf	1.073	—	805	—	805
Labor-ceiling ledgers	2x6	800	bf	.50	—	400	—	400
Labor-ceiling joists	2x6	7,000	bf	.394	—	2,758	—	2,758
Labor-ceiling blocking	2x6	1,600	bf	.50	—	800	—	800
Labor-wall sills, bolted	2x4	1,333	bf	.50	—	667	—	667
Labor-wall studding	2x4	10,666	bf	.358	—	3,818	—	3,818
Labor-wall blocking	2x4	1,333	bf	.50	—	667	—	667
Labor-wall top plates	2x4	2,666	bf	.429	—	1,144	—	1,144
Labor-wall bracing	1x4	112	bf	1.073	—	120	—	120
Labor-misc. stripping	1x3	400	bf	1.073	—	429	—	429
Labor-misc. grounds	1x2	200	bf	1.662	—	332	—	332
Labor-plywood roof sheathing	⅝"	10,240	sf	.312	—	3,195	—	3,195
Labor-plywood wall sheathing	½"	2,560	sf	.332	—	850	—	850
Labor-plywood floor sheathing	⅝"	10,240	sf	.312	—	3,195	—	3,195
Labor-plywood shear panels	½"	1,920	sf	.351	—	674	—	674
Equipment hoisting (48 hrs @ 70/hr)		70,000	bf	—	.048	—	3,360	3,360
					Total	28,538	44,290	72,828

Figure 2.23e

2.24
Finish carpentry

Finish carpentry is generally that which is nonstructural, exposed to view, and requires fine workmanship. Finish carpentry includes such items as:

> doors and door frames
> finish hardware installation
> cabinets and shelving
> milled trim
> non-milled but exposed to view trim
> wall paneling

Most of the finish carpentry materials are furnished by a millwork subcontractor for a lump sum price; but long before quotations are received from millwork companies, the estimator must anticipate the probable state of completion (fabrication) of the items upon their arrival at the jobsite. The general rules are:

1. Items which require shop fabrication, such as cabinets, doors, and door frames, will arrive complete and ready for installation.
2. Items which do not require shop fabrication, such as trim and paneling, will be loose or in bundles.
3. Items such as built-up shelving will be in pieces, ready to assemble.

The estimator further considers:

1. Everything requiring field labor.
2. Unloading of materials from trucks, handling, temporary storage and protection.
3. Materials not furnished by the millwork supplier.
4. Rough hardware and its cost, including such items as nails, bolts and screws.
5. Special tools, equipment, and scaffolding.
6. Handling and hoisting materials from storage to final positions.
7. Incidental associated items of work such as backing in partitions for the securing of cabinets and trim.

The take-off usually begins with doors, door frames, wood windows and wood window frames; next, loose trim and wall paneling; then cabinets and shelves; finally, miscellaneous items both inside and outside the building, such as stair handrailing, facia, etc.

The cost to install door frames in Figure 2.24a is used in examples in this book:

The unit costs to install millwork, in Figures 2.24a through 2.24i, are for use in the examples of finish carpentry pricing out given later in this book.

Door frames, wood or metal, (without casings) vary in cost with the following conditions:

1. Size of frame
2. Thickness of wall
3. Wall material (concrete, steel, wood)
4. Number and similarity of frames
5. Hardness of the wood frame and type of finish (natural, paint)

Unit labor costs in Figure 2.24a are based upon average conditions.

Doors vary in cost to install (hang) with the following conditions:

1. Size of door (width, height and thickness).
2. Construction of door (HC, SC, panel).
3. Quality of door (paint or natural finish).
4. Complexity (plain slab or ornamental).

5. Quantity and similarity of sizes.

The tables of unit labor costs shown in Figure 2.24b and Figure 2.24c are based upon average conditions, ornamental and solid core 1 3/4" thick and hollow core 1 3/8" thick, and include only the minimum hardware such as hinges and locks.

Figure 2.24a

Cost to Install Door Frames							
Opening Sizes							
Wall Material	2'-0" x 7'-0"	2'-6" x 7'-0"	2'-8" x 7'-0"	3'-0" x 7'-0"	3'-6" x 7'-0"	4'-0" x 7'-0"	5'-0" x 7'-0"
Concrete	20.00	21.00	22.05	23.15	24.30	25.52	26.80
Steel stud	19.04	20.00	21.00	22.05	23.15	24.30	25.52
Wood stud	18.13	19.04	20.00	21.00	22.05	23.15	24.30
Note: Add 10% per linear foot of height over 7'-0"							

Figure 2.24b

Cost to Hang Wood Doors							
Opening Sizes							
Type of Door	2'-0" x 7'-0"	2'-6" x 7'-0"	2'-8" x 7'-0"	3'-0" x 7'-0"	3'-6" x 7'-0"	4'-0" x 7'-0"	5'-0" x 7'-0"
Ornamental	30.60	32.13	33.74	35.43	37.20	39.06	41.00
Solid core (SC)	29.14	30.60	32.13	33.74	35.43	37.20	39.06
Hollow core (HC)	27.75	29.14	30.60	32.13	33.74	35.43	37.20
Note: Add 10% for natural finish doors (NF) Add $5.00/lf for doors over 7'-0" high							

Figure 2.24c

Cost to Hang Hollow Metal Doors							
Opening Sizes							
Type of Door	2'-0" x 7'-0"	2'-6" x 7'-0"	2'-8" x 7'-0"	3'-0" x 7'-0"	3'-6" x 7'-0"	4'-0" x 7'-0"	5'-0" x 7'-0"
Typical HM doors	21.76	22.85	24.00	25.20	26.46	27.78	29.17
Note: Add $5.00/lf for doors over 7'-0" high							

Finish hardware is customarily furnished by a hardware supplier by lump sum purchase order. The estimator calculates only the labor to install those special items which are not included in the basic door hanging unit prices. The costs vary widely by styles and conditions, but for purposes of this book, some typical hardware unit labor prices are given in Figure 2.24d.

Figure 2.24d

Labor to Install Finish Hardware	
Thresholds	18.13
Weatherstripping	36.26
Overhead closers	54.39
Panic devices	72.52
Push/pull plates	9.07
Kick plates	13.60

Cabinets and casework vary in cost to install according to the following conditions:

1. Quantity and similarity
2. Size of units
3. Quality (economy, custom, or premium)
4. Type of wood (soft to hard)
5. Type of finish (paint or natural)
6. Manner of fastening to walls or floors
7. Type of cabinet (base, wall, overhead, island)
8. Complexity (doors, drawers, shelving)
9. Material of counter top (laminated plastic, metal, wood)

Figure 2.24e shows unit labor costs typical of average conditions, used in the pricing out examples in this book. Base cabinets are priced by the linear foot, wall and overhead cabinets are priced by the square foot of front (face) area.

Figure 2.24e

Labor Cost to Install Cabinets			
	Economy Grade	Standard Grade	Premium Grade
Base cabinets with tops	4.67/lf	9.10/lf	12.14/lf
Wall cabinets	1.30/sf	1.95/sf	2.60/sf
Overhead cabinets	1.95/sf	2.93/sf	3.90/sf
Island cabinets (base)	4.55/lf	6.83/lf	9.10/lf
Open wall cabinets	.65/sf	.98/sf	1.30/sf

Shelving may be single and supported on brackets, assembled units ready to install, or material shop cut ready for assembly at jobsite. The following table, Figure 2.24f, of unit labor costs covers typical shelving conditions used in the pricing out examples later in this book.

Figure 2.24f

Labor Cost to Install Shelving			
Depth Of Shelf	Single Shelf On Brackets	Shelf Units Ready To Install	Loose Material For Jobsite Assembly
Inches	Per lf	Per sf	Per lf
12	2.25	.85	3.29
14	2.50	.91	3.46
16	2.77	.98	3.64
18	3.08	1.04	3.82
20	3.42	1.11	4.00
24	3.80	1.17	4.21

Milled trim labor costs vary according to the following conditions:

1. Dimension of the material
2. Hardness of the material
3. Length of typical pieces (shorter pieces cost more to install)
4. Number of joints, corners, etc.
5. Finish of wood (natural or painted)
6. Location (difficulty or ease of access)

The labor unit costs in Figure 2.24g are based on average conditions and will apply to examples later in this book.

Figure 2.24g

Labor Cost to Install Milled Trim Per lf						
	Hardwood			Softwood		
	Complexity			Complexity		
	A	B	C	A	B	C
Base trim	.51	.64	.75	.23	.29	.34
Window/door casings	.59	.74	.88	.26	.33	.39
Cornice mould	.73	.90	1.11	.33	.41	.49
Chair rail	.64	.80	.95	.30	.38	.43
Stair rail cap	1.46	1.82	2.22	.65	.81	1.00

Non-milled trim such as dimension lumber, clear, vertical grain, kiln-dried, etc. is priced out for labor similar to milled trim. Figure 2.24h is based on average conditions.

Figure 2.24h

Labor Cost to Install Non-Milled Trim Per lf				
	Hardwood		Softwood	
	Long Pcs Few Joints	Short Pcs Many Joints	Long Pcs Few Joints	Short Pcs Many Joints
1 x 2	.30	.38	.18	.23
1 x 3	.33	.40	.21	.26
1 x 4	.38	.47	.25	.30
1 x 6	.46	.57	.30	.36
2 x 4	.56	.70	.36	.44
2 x 6	.66	.83	.44	.53
2 x 8	.79	1.00	.53	.64

Wall paneling installation costs vary with the following conditions:

1. Application: nail to studs or stripping
2. Application: glue to gypsum board or other surface
3. Size of areas to be paneled (large or small)
4. Amount of cutting and fitting (openings, etc.)

The following labor unit costs, Figure 2.24i, are based upon average conditions.

Figure 2.24i

Labor Cost to Install Wall Paneling Per Square Foot		
Hardboard	⅛" thick	.55
Hardboard	¼" thick	.61
Softwood plywood, unfinished	¼" thick	.72
Hardwood plywood, unfinished	¼" thick	.78
Hardwood plywood, prefinished	¼" thick	.91
Hardwood plywood, prefinished	¾" thick	1.04
Boarding, softwood	¾" thick	.98
Boarding, hardwood	¾" thick	1.17

Rough hardware, such as nails and screws, is usually a lump sum item based upon a percentage of the total finish carpentry estimate.

The following Figure 2.24j is an example of a finish carpentry price-out, incorporating typical items found in commercial buildings.

Example of Finish Carpentry Price-Out

	Quantity		l	L	M	T
Door frames-wood in wood stud walls	2'-8" x 7'	36 ea	20.00	720	*	720
Door frames-wood in wood stud walls	3'-0" x 7'	8 ea	21.00	168	*	168
Door frames-wood in concrete walls	3'-0" x 7'	2 ea	23.15	46	*	46
Door frames-metal in concrete walls	3'-0" x 8'	4 ea	25.47	102	*	102
Doors-wood HC	2'-8" x 7'	36 ea	30.60	1,102	*	1,102
Doors-wood SC	3'-6" x 7'	8 ea	33.74	270	*	270
Doors-wood ornamental	3'-0" x 7'	2 ea	37.20	74	*	74
Doors-hollow metal	3'-0" x 8'	4 ea	30.20	121	*	121
Hardware-thresholds		6 ea	18.13	109	*	109
Hardware-weatherstripping		6 ea	36.26	218	*	218
Hardware-closers		12 ea	54.39	653	*	653
Hardware-panic devices		4 ea	72.52	290	*	290
Hardware-push-pull plates		14 ea	9.07	127	*	127
Hardware-kickplates		7 ea	13.60	95	*	95
Cabinets-base type standard grade		120 lf	9.10	1,092	*	1,092
Cabinets-base type premium grade		30 lf	12.14	364	*	364
Cabinets-wall type standard grade		420 sf	1.95	819	*	819
Cabinets-overhead standard grade		150 sf	2.93	440	*	440
Shelving 14" deep-KD		300 lf	3.46	1,038	*	1,038
Milled trim-base mold (cond C)		800 lf	.75	600	*	600
Milled trim-door casing (cond B)		1,564 lf	.74	1,157	*	1,157
Non-milled trim 1x4 hardwood		1,000 lf	.47	470	*	470
Non-milled trim 1x6 softwood		960 lf	.30	288	*	288
Wall paneling ¼" prefinished hardwood		480 sf	.91	437	*	437
Rough hardware (2% of total cost)		ls	—	—	264	264
Total				10,800	264	11,064

*Materials are to be furnished by suppliers and the cost will be listed separately on the bid sheet.

Figure 2.24j

The miscellaneous metal for a project is usually furnished by a subcontractor for a lump sum price. Some items are installed by the subcontractor and other items are delivered to the jobsite for installation by the general contractor's employees. Through his knowledge of the Customs-of-the-Trades (see Section 1.21—Customs-of-the-Trades) and his conversations with miscellaneous metal subcontractors, the estimator anticipates the items which will be furnished only, and he estimates the cost to install them.

Customarily the miscellaneous metal subcontractor furnishes and installs the following typical items:

1. All nonstructural metal (excluding sheet metal) requiring the special skills of metal workers for their assembly and erection.
2. Items which may be installed more or less independently of other trades.
3. Items of metal which attach to other metal.

Customarily the miscellaneous metal subcontractor furnishes only (does not install) the following typical items:

1. Bolts, sleeves, angle or bar thresholds, stair nosings, inserts, etc., which are to be embedded in concrete.
2. Timber connectors such as custom-made (non-stock) straps, joist hangers, etc., which require installation by carpenters.

These are items which require no special metal worker skills and are more convenient for other trades to install.

The take-off for installation of miscellaneous metal serves also as a check sheet to analyze and compare sub-bids. Figure 2.25a, includes all miscellaneous metal items in a hypothetical project. Because of the diversity of installation conditions, no tables of costs are given here; the estimator must use his best judgment and experience.

Example of a Miscellaneous Metal Labor Price-Out				
	Quant	**Production**	**Unit Cost**	**Total**
Braces	—	sub install	—	—
Tie straps	1 ea	1.5 mhr	27.20	27
Angle door headers	—	sub install	—	—
Ladder to attic	3 ea	2.0 mhr	36.26	109
Steel handrail	—	sub install	—	—
Bent plates at vents	24 ea	.75 mhr	13.60	326
Bike racks	6 ea	1.75 mhr	31.73	190
Ledger angles	1,600 lf	25 lf/hr	.73	1,168
Stair nosings	500 lf	10 lf/hr	1.81	905
Catch basin cover	4 ea	.75 mhr	13.60	54
Pipe guard posts	13 ea	.5 mhr	9.07	118
Bench supports	4 ea	.75 mhr	13.60	54
Sleeves for railings	65 ea	.25 mhr	4.53	295
Aluminum sunscreen	—	sub install	—	—
Burglar bars & frames	2 ea	1.5 mhr	27.20	54
Channel door frames	5 ea	1.5 mhr	27.20	136
Angle thresholds	8 ea	.5 mhr	9.07	73
Angle at dock edge	96 lf	25 lf/hr	.73	70
Trench cover & frame	120 sf	20 sf/hr	.91	109
Steel stairs	—	sub install	—	—
Roof hatch	—	sub install	—	—
Total				3,688

Figure 2.25a

2.26
Miscellaneous trades

Section 1.7—Divisions of the Estimate names two main divisions: I-General Contractor Work, and II-Subcontractor Work. Most of the construction trades clearly fall under one or the other division. There are, however, a few trades which fall between those two divisions (see Section 2.11—Subcontract Work) and are figured by the estimator just in case no realistic sub-bids are obtained. Examples: caulking and sealing, concrete curbs and sidewalks, concrete structures for mechanical trades, patching of asphalt concrete paving.

These are not budgets (see Section 3.1—Budgeting Subtrades), but genuine estimates, carefully taken off and priced out. They are described below.

Caulking and sealing materials and methods are usually described in the specifications and in the drawings. Customs-of-the-Trades place certain caulking items under subcontract work (sheet metal, roofing, mechanical, glass and windows, electrical, etc.), so the estimator takes off only those items which are normally done by the general contractor's employees. Some typical caulking and sealing items are:

1. Exterior door frames
2. Window frames (when not a subcontract item)
3. Wood siding and trim
4. Exterior plaster joints
5. Interior plaster and gypsum board joints
6. Joints in tilt-up concrete wall panels
7. Control joints in masonry

The cost of caulking and sealing depends upon:

1. The type of materials used (elastomeric, acrylic, etc.)
2. The cross section size (1/2" x 1/2"; 1/2" x 3/4", etc.)
3. Location and difficulty of application (scaffolding cost is in addition to the unit costs given in Figure 2.26c)

Figures 2.26a through 2.26d will be used in the pricing out examples given later in this book.

Quantity of Caulking Material Required For Various Sizes of Joints

Size of Joint	Linear Feet Per Gallon
¼" x ¼"	388
¼" x ½"	145
½" x ½"	77
½" x ¾"	56
¾" x ¾"	34
¾" x 1"	27
1" x 1"	19

Figure 2.26a

Backer Rods

Size	Labor	Material	Total
¼"	.36	.18	.54
½"	.38	.28	.66
¾"	.39	.52	.91
1"	.42	.75	1.17

Figure 2.26b

143

Figure 2.26c

	Cost of Caulking and Sealing							
	A		**B**		**C**		**D**	
Size	Elastomeric		Oil Base		Acrylic Base		Butyl	
	l	m	l	m	l	m	l	m
¼″ x ¼″	.46	.03	.52	.04	.55	.05	.59	.05
¼″ x ½″	.52	.07	.56	.05	.59	.08	.62	.08
½″ x ½″	.56	.13	.60	.12	.62	.16	.70	.16
½″ x ¾″	.60	.18	.62	.20	.70	.21	.78	.22
¾″ x ¾″	.62	.27	.65	.26	.78	.35	.83	.36
¾″ x 1″	.65	.36	.75	.34	.83	.44	.88	.46
1″ x 1″	.75	.52	.83	.46	.88	.62	.94	.65

	E		**F**		**G**		**H**	
	Latex		Polysulphide		Polyurethane		Silicon	
	l	m	l	m	l	m	l	m
¼″ x ¼″	.47	.03	1.04	.10	.65	.08	1.11	.12
¼″ x ½″	.46	.07	1.12	.23	.72	.17	1.17	.29
½″ x ½″	.52	.12	1.20	.44	.78	.33	1.24	.55
½″ x ¾″	.56	.16	1.25	.61	.85	.44	1.30	.74
¾″ x ¾″	.59	.23	1.30	1.00	.91	.73	1.43	1.22
¾″ x 1″	.63	.33	1.43	1.25	.98	.91	1.56	1.55
1″ x 1″	.68	.49	1.56	1.78	1.04	1.30	1.69	2.18

The following, is an example of a typical caulking estimate. For convenience, capital letters (H) etc.,indicate the materials from Tables 2.26b and 2.26c.

Figure 2.26d

Example of a Caulking and Sealing Price-Out								
			Quant		l	m	L	M	T
Door frames	¼″x½″	(H)	240	lf	1.17	.29	281	70	351
Wood trim	¼″x¼″	(H)	720	lf	1.11	.12	799	86	885
Exter plaster	¼″x½″	(F)	450	lf	1.12	.23	504	104	608
Backer rod	½″		450	lf	.38	.28	171	126	297
Inter gyp bd	½″x½″	(A)	1,100	lf	.56	.13	616	143	759
Tilt-up conc	½″x¾″	(H)	960	lf	1.30	.74	1,248	710	1,958
Backer rod	¾″		960	lf	.39	.52	374	499	873
Total							3,993	1,738	5,731

Concrete curbs and sidewalks-see Section 3.1—Budgeting Subtrades.

Structures for mechanical trades-see Section 2.21—Cast-In-Place Concrete, Figures 2.21x and 2.21y.

Patching asphalt paving over mechanical trenches, along new curbs, or wherever existing asphalt has been disturbed, is an item which is not always included by subcontractors in their bids; therefore it is figured by the general estimator, and the cost depends upon the following conditions:

1. Quantity (the cost for mobilization is nearly the same for small quantities as large).
2. Accessability for machinery (otherwise, increased handwork is required.)
3. Thickness.
4. Base course, if any, and thickness.
5. Continuity (long stretches or small pieces.)

Figure 2.26e displays average unit costs which will be used in the pricing out examples later in this book.

Figure 2.26e

Total Cost of Aggregate Base Course, Labor, Mat'l. and Equipment Cost Per Square Foot					
	4" thk	5" thk	6" thk	7" thk	8" thk
Under 1000 sf	.40	.48	.56	.64	.73
1000 to 2000 sf	.38	.46	.53	.61	.70
2000 to 5000 sf	.35	.43	.51	.59	.68
Over 5000 sf	.33	.40	.48	.56	.65

Figure 2.26f

Asphalt Concrete (No Base), Labor, Material, and Equipment Cost Per Square Foot					
	1½"	2"	2½"	3"	3½"	4"
Under 1000 sf	1.34	1.70	1.94	2.05	2.37	2.73
1000 to 2000 sf	1.12	1.42	1.61	1.72	1.98	2.28
2000 to 5000 sf	.94	1.18	1.34	1.43	1.65	1.90
Over 5000 sf	.78	.99	1.12	1.20	1.38	1.59

Figure 2.26g

Example of Estimated Cost For Asphalt Concrete Patching
Cost Per Square Foot

2" thick w/no base	6,200 sf @ .99 = 6,138
3" thick w/6" base	3,000 sf @ 1.94 = 5,820
1½" thick w/4" base	900 sf @ 1.74 = 1,566
TOTAL	13,524

2.27 Fringe benefits payroll taxes

In all the previous examples in this book, labor was based on the payroll level and no consideration was given to the indirect labor costs (fringe benefits, taxes and insurance). These indirect costs may be incorporated into the estimate at one or two later points:

1. They may be added in one lump sum to the total cost of each trade such as demo, concrete, carpentry, etc.
2. They may be added to the grand total of all labor in the project.

The first method is preferable because it completes the estimated cost of each trade, for comparison to possible sub-bids. In all of the examples that follow this section, these indirect costs (abbreviated F.B.) will be added to the end of each trade.

Figure 2.27a is an example of a general contractor's labor costs. The table requires adjusting each time one of the trades receives an increase in base pay or fringe benefits.

As Figure 2.27a shows, the different trades receive not only different base (take-home) pay amounts, but the contractors pay different amounts of fringe benefits to various agencies on the worker's behalf. It would be impractical for the estimator to apply fringe benefits to each trade separately. Here are two convenient ways to calculate this cost:

1. If the total labor at base pay level is, say 235,000 and the estimator judges trades in the proportions of: carpenters ½, cement masons ¼, and laborers ¼, using FB's in Figure 2.27a

carpenters	235,000 x .50 x .41 =	48,175
cement masons	235,000 x .25 x .46 =	27,025
laborers	235,000 x .25 x .53 =	31,138
		106,338

$$\frac{106,338}{235,000} = .453 \text{ or } 45.3\%$$

2. Instead of doing the calculations above, use the 46% proposed in Figure 2.27a as an approximate average for the trades, thus:

$$235,000 \times .46 = \qquad 108,100$$

	Carpenter	Cement Mason	Laborer
Base Pay (Payroll)	**18.13/hr**	**17.29/hr**	**15.10/hr**
All union benefits	2.90	3.92	4.17
Worker's comp insurance	2.27	1.90	1.82
Soc Sec & unempl insurance	1.92	1.80	1.72
General liability	.34	.33	.29
	25.56/hr	25.24/hr	23.10/hr
% increase over payroll	41%	46%	53%
For estimating purposes, use average of 46%			
Foreman - add to base pay	1.25/hr	1.00/hr	.75/hr

Figure 2.27a

Accuracy in the estimating of fringe benefits is important, considering the large sums extended, but unless a project calls for an overwhelming majority of one of the trades, the average, as in Example #2, is the practical approach.

Part 3
Concluding Considerations

3.1 Budgeting subtrades

A budget is simply a rough estimate. It lies somewhere between a true estimate and a mere plug-in-figure (see Section 3.4—Plug-in-Figures). A true, bona fide, estimate involves a finely detailed take-off and price-out by an estimator experienced in the trade. In the masonry trade, for instance, the number of pieces of each type of block, cubic feet of mortar, pounds of rebar, quantities of accessories, labor hours and equipment hours are taken into account. But a budget may be based on nothing more than the square feet of wall surface. The result is a "ballpark" figure. Yet, no matter how well figured a trade may be, if the intention is to sublet, the amount figured is only a budget. That is because a contractor cannot properly estimate a cost, or guess a markup, for another contractor.

The purposes of a budget are:

1. To provide a means of analyzing a subtrade for intelligent discussion with prospective bidders.
2. To isolate items of potential dispute, or those which could easily be missed by sub-bidders.
3. To check and validate a single bid (when there are no other bids).
4. To have a figure in reserve against the possibility of not receiving a sub-bid, or in case of subcontractor default.
5. For follow-up negotiations of a subcontract.
6. To provide an alternative when there is not time nor sufficient information to make a proper estimate.
7. To complete a cost item when a technical service quotes only unit prices.

The estimator who energetically and consistently procures sub-bids in all trades seldom needs to budget any of them. The need to budget increases for unusual projects, trades, and projects which are located beyond the range of local subcontractors; dependence upon unknown subcontractors may be cause for budgeting.

A personal reason for budgeting is the continued education of the estimator and the broadening and sharpening his skills.

A very important point to remember is that an estimated amount is not necessarily the same as a bid amount. In making a budget, the estimator may try:

1. To ascertain the approximate amount a portion of a project would cost if the general contractor accomplished it by hiring workmen and

purchasing the materials.

2. To establish an amount which would be acceptable to a subcontractor through later negotiations.

3. To strike a compromise between the above two cost levels.

Budgets are more dependable when figured at the bare cost level and a suitable subcontractor markup added. Under uncertain, or unknown workload conditions, it sometimes happens that a subcontractor is unwilling to commit himself to work a great many months in the future, except at a very high markup amount. A general contractor's budget, compared to a sub-bid, may indicate that a noncompetitive case exists and provide cause for rejection of the subcontractor's bid.

Following are some typical trades and methods of budgeting their costs. For current unit costs of many trade items additional to those mentioned in this section, the reader is referred to R.S. Means *Building Construction Cost Data*.

A. Demolition (site clearing)

Typical obstacles (to the new construction project) which are customarily removed by demolition subcontractors include: trees, shrubbery, fences, curbs, walks, paving and buildings. "Structural Demolition", incidental to remodeling, repair or alteration is included in Section 2.20—Demolition.

Generally the line of separation is this: Items which can be demolished rapidly and by machinery are subcontractor's work, and will be described in this section; items which require careful measurements and hand labor are general contractor's work, and are described in Section 2.20.

Figure 3.1a is an example and explanation of a demolition estimate.

Sample Budget of Sitework Demolition		Q	e	T	Ref
1. Wood frame building-2 story		8,000 sf	.92	4,480	2.20n
Susp clg met & gyp bd		12,000 sf	.235	2,820	2.21l
Floor framing & finishes		9,000 sf	.25	2,250	2.20n
Part'ns, wood & gyp bd		9,600 sf	.365	3,504	2.20m
Conc floor slab on grd 4"		8,000 sf	.38	3,040	2.20f
Concrete footings		120 cy	79.00	9,480	2.20i
2. Trees & stumps	12"	2 ea	105.30	211	2.20k
3. A.C. Paving-saw cut	1½"	600 lf	.45	270	2.20b
remove	3"	18,000 sf	.12	2,160	2.20j
4. Concrete paving	6"	16,000 sf	.33	5,280	2.20g
5. Concrete std gutter curb		200 lf	1.62	324	2.20h
6. Concrete manholes	9' deep	450 cf	3.40	1,530	2.20o
7. Chainlink fence	6' high	320 lf	1.30	416	2.20o
Bare cost subtotal				38,645	
Subcontractor markup 20%				7,729	
Total budget				**46,374**	

Figure 3.1Aa

The explanation of this budget is as follows:

1. The 2-story wood frame building has a ground floor area of 8,000 square feet and a second floor area of 8,000 square feet. If construction time permits, the cost may be estimated on the basis of dismantling and salvaging the materials, but these unit prices cover

complete wrecking and hauling to the nearest legal dumpsite.

2. The removal of trees and stumps is priced directly from Figure 2.20k.
3. The removal of asphalt pavement often requires sawcutting where a portion of the pavement is to remain. To break out the A C, load on trucks and haul to a dump site approximately three miles away, see Figure 2.20j.
4. The cost to remove concrete pavement depends upon the following conditions:
    a. Thickness of concrete
    b. Whether or not reinforced
    c. Hardness of concrete
    d. Distance of haul to dump

In this example, and Figure 2.20g, average conditions are presumed.

5. Concrete curbs of quantities large enough to be broken up by machinery, loaded on trucks and hauled to a dump approximately three miles distant may be budgeted by unit costs similar to those in Figure 2.20h.
6. The cost to remove manholes depends upon the size, particularly the depth. See Figure 2.20o.
7. Chain link fencing and many other miscellaneous items are budgeted by reference to cost records available to the estimator. See Figure 2.20o.

The subtotal is presumed to be a subcontractor's bare cost. Since the general (prime) contractor may prefer to sublet this work, to make a realistic comparison to any possible sub-bids the estimator adds a markup which he judges to be typical for the subtrade.

The total budget is intended to be a correct amount whether the work will be accomplished by the general contractor or by a subcontractor.

B. Earthwork (site grading)

A jobsite visit often produces valuable information not indicated on the drawings. A good take-off is necessary to produce a dependable budget. While taking off the quantities, the estimator visualizes the type and number of pieces of equipment most suitable to the loosening, moving, spreading, compacting and finishing of the earthen materials.

The estimator must decide the amount of time to spend in the take-off commensurate with the circumstances of the project, the conditions of the bidding, and the time available to do the budgeting.

Typically there are four grade levels, as depicted in Figure 3.1Ba, to be identified.

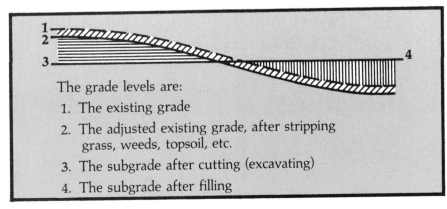

The grade levels are:

1. The existing grade
2. The adjusted existing grade, after stripping grass, weeds, topsoil, etc.
3. The subgrade after cutting (excavating)
4. The subgrade after filling

Figure 3.1Ba

As the quantities are taken off, analysis goes typically as follows:

1. An itemized list for clearing
   a. How can items be broken or cut up?
      What kind of equipment is most suitable?
   b. Where can the debris be disposed of?
2. Stripping
   a. Grass, weeds, brush, etc. Does any exist?
   b. Topsoil. Does any exist? What is to be done with it?
   c. Look for material unsuitable for use in the project; silt, clay, cobbles, etc. What is to be done with it?
   d. What equipment is required for stripping and hauling?
   e. Where can the debris be disposed of?
3. Cubic yards of earth material to be excavated.
   a. Stockpile for future use?
   b. Push to nearby fill areas?
   c. Load and haul to nearby off site fill areas?
   d. Load and haul to distant dump site?
   e. What type of equipment is the most suitable?
4. Cubic yards of earth material for fill.
   a. Use material excavated on site?
   b. Import material from off site?
   c. Spread and compact the fill materials?
   d. What degree of compaction?
   e. What type of equipment is the most suitable?
5. Square feet of areas to receive different types of finishes.
   a. Fine grade to within 1/10th of a foot, plus or minus, of exact elevations.

Questions regarding location of dumps and borrow areas, suitability of existing materials for fill, required imported select materials, degree of compaction required, and so forth, are answered variously as follows:

1. As described in the specifications.
2. From visual inspection of the jobsite and surrounding vicinity.
3. From study of data from test pits, borings and geological information.
4. From material suppliers and other knowledgeable persons.

The quantities as first taken off are in their natural (bank) state of compaction. When loosened up by the excavation equipment, the material increases in volume and since excavators, loaders and trucks are limited to their design capacities, "Loose yards" are the real quantities to be used in excavation cost budgeting. To convert bank yards to loose yards, apply appropriate percentages, a few of which are suggested in Figure 3.1Bb.

Figure 3.1Bb

Volume Increase When Bank Yards are Loosened by Excavation	
Crushed rock or gravel	15%
Sand	20%
Silt	22%
Loam (average soil)	24%
Decomposed granite	25%
Clay	26%

After the quantities are determined, intermediate and special operations are listed, specifications are studied and special details are noted. These include:

1. Scarifying and compacting of native soil.
2. Benching of slopes.
3. Cutting for curbs, walks, driveways, slabs or streets.
4. Cutting and shaping ditches.
5. Special contouring (mounds, etc.)
6. Importing of special material.
7. Mass excavation of open pits.
8. Spreading of topsoil.
9. Fine grading.
10. Dewatering.

Any two projects having the same area in square feet will cost approximately the same for stripping, scarifying, compacting and fine grading. The main difference in cost will depend upon such variables as:

1. Hardness of the material.
2. Looseness, dryness, wetness.
3. Thickness of stripping required.
4. Depth and degree of scarifying and compacting required.
5. Accessibility of site, and intricacy of fine grading requirements.

The major difference in cost between projects of the same surface area is in the volume of earth to be moved (depth of cut and fill).

The size (capacity) of a piece of equipment is important when:

1. Unusual power is needed.
2. Unusual reach (digging depth) is required.
3. Speed is essential.

A large excavator will not reduce the cost unless it can increase the production relative to rental/operating cost. As an example, in soft earth:

$$\text{a } 55/\text{hr excavator moving } 55 \text{ cy/hr}=1.00/\text{cy}$$
$$\text{a } 110/\text{hr excavator moving } 110 \text{ cy/hr}=1.00/\text{cy}$$

Yet, the larger machine decreases the overall time and thus produces a form of saving.

A greater difference may result if the earth is hard:

$$\text{a } 55/\text{hr excavator moving } 25 \text{ cy/hr}=2.20/\text{cy}$$
$$\text{a } 110/\text{hr excavator moving } 60 \text{ cy/hr}=1.83/\text{cy}$$

The greater power of the larger machine can increase the relative production; but the smaller machine may be more maneuverable and therefore, a better choice in restricted areas.

These considerations are very important when actually made in the field, but are only theoretical in cost budgeting.

From the foregoing, the main qualifications of the estimator necessary for making dependable excavation budgets are:

1. Skill at quantity surveying.
2. The ability to analyze the operations and plan the numbers and types of suitable equipment.
3. Access to current equipment rental rates.
4. Knowledge of production capacities of equipment handling different materials under various conditions.

Hardness is only one of the factors that affect excavation production.

Others are roughness, moistness, looseness, rockiness and room to maneuver. A judgment must be made. Even soft, ideal material may, in certain circumstances, be slow to move. Figure 3.1Bc is a scale used in the pricing out examples given later in this book.

Figure 3.1Bc

Earth Moving Difficulty Factors

Cemented soils	.50
Dense soils	.63
Firm soils	.83
Average soils	1.00
Loam	1.12
Decomposed granite (loose)	1.25
Topsoil	1.30
Sand (loose and dry)	1.43

Notice that "average" soils is not a mean. The production capacities of machinery limit the quantity of earth that can be moved under even the best conditions.

Two examples of rough excavation budgets are given hereafter. The first is a relatively small project; the second is large enough for the use of high-production equipment such as rubber-tired self-loading scrapers.

Example #1 per sketches Figure 3.1Bd through 3.1Bj:

The take-off gives:

1. Strip 6" and haul three miles	890	cy
2. Scarify and compact	20,000	sf
3. Cut, and fill on site	1,280	cy
4. Excavate & haul three miles	1,800	cy
5. Import and compact in place	1,000	cy
6. Fine grade overall	40,000	sf

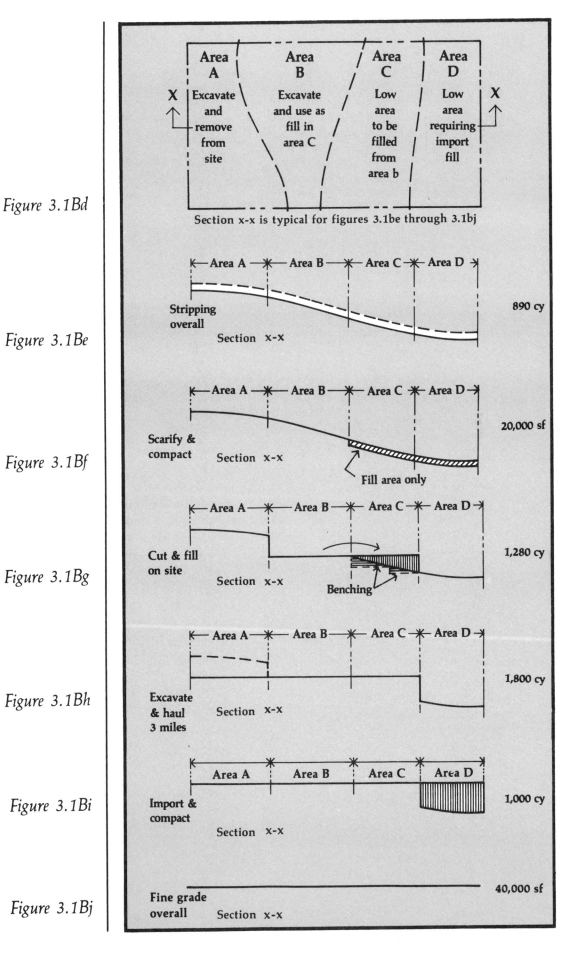

Figure 3.1Bd

Area A — Excavate and remove from site
Area B — Excavate and use as fill in area C
Area C — Low area to be filled from area b
Area D — Low area requiring import fill

X ↑ X ↑

Section x-x is typical for figures 3.1be through 3.1bj

Figure 3.1Be

Area A — Area B — Area C — Area D

Stripping overall   Section x-x   890 cy

Figure 3.1Bf

Area A — Area B — Area C — Area D

Scarify & compact   Section x-x   20,000 sf

Fill area only

Figure 3.1Bg

Area A — Area B — Area C — Area D

Cut & fill on site   Section x-x   1,280 cy

Benching

Figure 3.1Bh

Area A — Area B — Area C — Area D

Excavate & haul 3 miles   Section x-x   1,800 cy

Figure 3.1Bi

Area A — Area B — Area C — Area D

Import & compact   Section x-x   1,000 cy

Figure 3.1Bj

Fine grade overall   Section x-x   40,000 sf

153

Budget Cost

1. Strip and waste to three miles	890	cy	2.40	2,136
2. Scarify and compact fill areas	20,000	sf	.05	1,000
3. Cut and fill on site	1,280	cy	3.17	4,058
4. Exc and haul 3 miles to dump	1,800	cy	3.05	5,490
5. Import-purchase FOB Jobsite	1,000	cy	5.74	5,740
spread and compact	1,000	cy	.95	950
6. Fine grade overall	40,000	sf	.036	1,440

Subtotal	20,814
Sub Markup 20%	4,163
Total Budget	24,977

This budget is explained in numerical order as follows:

1. Assume: excavator/loader @ 75/hr = 75
   trucks 3 each @ 55/hr = 165
   Total 240/hr

   Production: say, 100 cy/hr, $\dfrac{240/hr}{100\ cy/hr} = 2.40/cy$

2. Before fill is placed in areas C and D, the stripped surface is scarified to 6" or more in depth and recompacted typically as follows:
   dozer/compactor 90
   water truck 60
   Total 150/hr

   Production: 3000 sf/hr, $\dfrac{150/hr}{3000\ sf/hr} = .05/sf$

3. The difficulty factor, Figure 3.1Bc, is judged to be .75 (25% slower to move than average). The distance between cut and fill areas is so short that loosened material may be pushed by dozer, or carried by loader, and
   dozer 80
   compactor 50
   water truck 60
   Total 190/hr

   Production: assume 2 days (16 hrs) $\dfrac{1,280\ cy}{16\ hr} = 80\ cy/hr$

   80 x .75 = 60 cy/hr $\dfrac{190/hr}{60\ cy/hr} = 3.17/cy$

4.   dozer 80
       loader 60
       trucks 3 ea @ 55 165
       305/hr

   Production: say, 100 cy/hr, $\dfrac{305/hr}{100\ cy/hr} = 3.05/cy$

5. A quotation of 4.25/ton is received, and the material weighs 2700 lb/cy;

   $\dfrac{2700\ lb/cy}{2000\ lb/ton} = 1.35\ tons/cy$

   4.25 x 1.35 = 5.74/cy

   To spread and compact, use the same equipment as in item #3 above - 190/hr

$$\frac{190/hr}{200\ cy/hr} = .95/cy$$

Production: say 200 cy/hr;

6. Equipment: Grader 65/hr
   Production: say 1800 sf/hr; $\dfrac{65/hr}{1800\ sf/hr} = .036/sf$

Example #2 per Figure 3.1Bk

The take-off gives:

1.	Remove AC paving, 3" thick to five miles	22,000 sf
2.	Remove concrete wall and footing to five miles	300 cy
3.	Remove trees, 12" diameter, & stumps to five miles	5 ea
4.	Strip & waste 4" deep to area E	6,000 cy
5.	Cut, and fill from area B to A	1,500 cy
6.	Cut & fill areas C to D	12,500 cy
7.	Cut & waste areas C to E	7,000 cy
8.	Scarify & compact areas A & D	12,000 sf
9.	Fine grade overall	450,000 sf

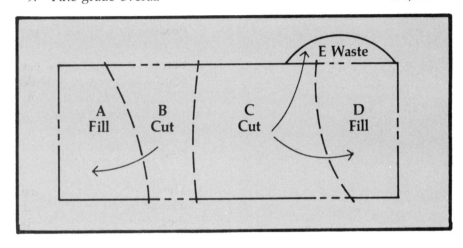

Figure 3.1Bk

Budget Cost

		Q		s	T
1.	Remove AC paving	22,000	sf	.13	2,860
2.	Remove conc wall & ftg	300	cy	80.10	24,030
3.	Remove trees & stumps	5	ea	116.30	582
4.	Strip & waste	6,000	cy	.59	3,540
5.	Cut, & fill on site	1,500	cy	2.03	3,045
6.	Cut, haul & compact	12,500	cy	1.34	16,750
7.	Cut, haul & waste	7,000	cy	.68	4,760
8.	Scarify & compact	12,000	sf	.05	600
9.	Fine grade	450,000	sf	.023	10,350

Subtotal	66,517
Subcont. Markup 20%	13,303
Total Budget	79,820

This budget is explained in numerical order as follows:

1. From Table 2.20j the cost for removal of AC paving and hauling three miles (one way) would be .12/square foot. To haul five miles increases the cost by truck travel time only—4 miles @ 20 miles per hour = 12 minutes = .2 hours.
   One load of 10 cubic yards = 1080 square feet of 3" pavement.
   One truck @ 55/hr x .2 = 11; $\dfrac{11}{1080}$ = .01; .12 + .01 = .13/sf

2. See Figures 2.21i and 2.20m: 300 cubic yards @ 79 = 23,700. Per item #1 above, add 11/truck load to increase haul distance from 3 to five miles;

$$\frac{11}{10\ cy} = 1.10/cy \text{ and } 79 + 1.10 = 80.10/cy$$

3. From Table 2.20k the cost to remove a 12" diameter tree with stump and haul three miles is 105.30 each. If one tree with stump makes a truck load, add 11.00 for extra mileage (as in #1 & #2 above), and 105.30 + 11.00 = 116.30/tree.

4. The topsoil surface layer has a difficulty factor of 1.3, per Figure 3.1Bc.

Equipment: self-loading 12 cy scraper:	110/hr.(Fig. 2.10a)	
Production: 12 cycles per hour;	12 x 12 = 144 cy/hr	
144 x 1.3 = 187 cy/hr;	110/187 = .59/cy	

5. The difficulty factor is 1.12

Equipment: dozer, 105 hp	60/hr
compactor	50/hr
water truck 1/2 time	26/hr
	136/hr

Production: 12 cycles x 5 cy	= 60 cy/hr
60 x 1.12 = 67 cy/hr;	136/67 = 2.03/cy

6. The difficulty factor is 1.12

Equipment: dozer 1/2 time	30/hr
scraper 12 cy	110/hr
compactor	50/hr
water truck 1/2 time	26/hr
	216/hr

Production: 12 cycles/hr; 12 x 12	= 144 cy/hr
144 x 1.12 = 161 cy/hr;	216/161 = 1.34/cy

7. In this example, the cost to cut and waste is the same as to strip and waste (item #4 above), except for the difference in the difficulty factor:

    144 x 1.12 = 161 cy/hr;   110/161 = .68/cy

8. Use the same unit cost as item #2 in example #1.
9. Equipment: grader   70/hour
   Production: 3,000 sf/hr: 70/3,000/sf = .023/sf

C. Concrete pavement

Pavement not placed by paving machinery may be figured similar to floor slabs. Typically the following steps are involved:

Layout and fine grading
Select base course
Edge forms
Setting of screeds
Construction, expansion and control joints
Concrete, placing and vibrating
Finishing
Curing and protection

Example:

12,000 sf of 6", 3,000 psi, 1 1/2 aggregate, on 6" thick class II base course, broom finish.

		Budget Estimate					
	Quant	l	m/e	L	M/E	T	Ref
1. Fine grade	12,000 sf	.07		840		840	2.21c
2. Base course	256 cy	1.48	17.95	379	4,595	4,974	2.21dd
3. Edge forms	660 lf	1.77	.79	1,168	521	1,689	2.21i
4. Set screeds	12,000 sf	.085	.028	1,020	336	1,356	2.21aa
5. Form keyed cj's	440 lf	1.76	.30	774	132	906	
6. Expan joints							
¾" x 5 ½"	220 lf	.468	.39	103	86	189	2.21ff
7. Conc & placing	240 cy	7.74	73.16	1,858	17,558	19,416	2.21s/t
8. Finish & cure	12,000 sf	.22	.022	2,640	264	2,904	2.21bb
9. Saw cut for jts	440 lf		1.06		466	466	2.20a
10. Joint sealant							
¾" x 1"	220 lf	.75	.34	165	75	240	2.26c
11. Joint sealant							
½" x ¾"	440 lf	.62	.20	273	88	361	2.26c
Total				9,220	24,121	33,341	

Explanation:

1. Cost to fine grade medium soil by hand, per Figure 2.21c.
2. Labor, per Figure 2.21dd, is 1.48; material and equipment, per Figure 2.21dd and 2.21ee come to: 13 + 4.95 = 17.95. Note that the quantity of 256 cubic yards includes a swelling factor of 15% in accordance with Figure 3.1Bd.
3. Assume 4 uses of 6" thick slab forms.
4. Screeds for 6" slab on rock aggregate base course.
5. Approximately the same as edge form cost.
6. See Figure 2.21ff.
7. Note that a waste factor of 8% is included in the quantity in accordance with Figure 2.21u. Consult Figures 2.21s and 2.21t for labor, material and equipment cost.
8. See Figure 2.21bb, broom finish.
9. See Figure 2.21a for 1" deep saw cut.
10. Find cost in Figure 2.26c under oil base.
11. Find cost in Figure 2.26c under oil base.

A quick, but less accurate way of budgeting the cost of concrete pavement is to use a table such as Figure 3.1Ca, and add for base course and other special features.

Cost per SF for Concrete Pavement (No Base Course)	
Thickness	**Cost per SF**
6"	2.26
8"	2.66
9"	2.70
10"	3.10
12"	3.60

Figure 3.1Ca

For the example previously given, use of Figure 3.1Ca would produce the following budget:

12,000 sf 6" conc pavement @ 2.26	=	27,120
250 cy base course @ 20.00	=	5,120
Total		32,240

D. Concrete sidewalks

Sidewalks may be budgeted in a manner similar to concrete pavement. Variables affecting the cost include:

1. Thickness of concrete and concrete strength.
2. Type of finish, color, etc.
3. Quantity in square feet, also proportion of edges to be formed.
4. Complexity (curves, etc.)

If a simple budget is desired, a table similar to Figure 3.1Da may be used. It assumes 2500 psi concrete, average edge forms (in linear feet, 30% of the total square feet of walks), 4" thickness, few curves, and standard sidewalk finish.

	Sidewalk Costs per SF			
	Under 1000 sf	1000 to 5000 sf	5000 to 10,000 sf	Over 10,000 sf
Sidewalk	2.08	1.90	1.75	1.69
Add rocksalt	.26	.25	.23	.22
Add color	.33	.29	.26	.23
Add 10% more forms	.04	.04	.04	.04

Figure 3.1Da

Example:

18,000 square feet sidewalks, 40% to receive saltrock finish, linear feet of edge forms are 50% of surface area (20% more than values in Figure 3.1Da).

Basic cost of sidewalk	1.69/sf
Rocksalt .22 x .40	.09/sf
Edge forms .04 x 2	.08/sf
	1.86/sf

18,000 x 1.86 = 33,480

E. Concrete curbs

Curbs may be standard or non-standard, plain, gutter, or roll-type. The two main variables affecting the cost of standard curbs are quantity and complexity (short lengths number of corners and curves) as shown in the following Figure 3.1Ea.

Cost of Concrete Curbs per LF				
(no excavation or backfill included)				
	Under 500 lf	500 to 1000 lf	1000 to 2000 lf	2000 lf & over
Gutter type-straight	9.24	8.86	8.39	7.97
moderate curves	9.69	9.22	8.78	8.39
many curves	10.19	9.70	9.24	8.80
Plain type-straight	8.49	8.09	7.70	7.33
moderate curves	8.88	8.53	8.14	7.72
many curves	9.39	8.94	8.53	8.11
Mow strip 6" x 6"				
straight	5.85	5.55	5.28	5.02
moderate curves	6.15	5.82	5.54	5.27
many curves	6.46	6.11	5.81	5.53
To excavate and backfill, add .90/lf				

Figure 3.1Ea

F.	For budgeting, see Figures 2.21x and 2.21y.
Catch-basins and man-holes	

G.	Non-standard items such as the following may be priced out by methods described in Section 2.21—Cast-In-Place Concrete.
Misc. site work concrete items	Planter curbs Cross gutters Steps on ground Headwalls Equipment bases Retaining walls

H.	The cost of reinforcing steel varies with the size of bars, number of cuts and bends and difficulty of placing.
Reinforcing steel and mesh	Bars are taken off in linear feet and converted to pounds by the factors in Figure 3.1Ha.

Weights of Rebar in Pounds per lf

#3	.376
#4	.668
#5	1.043
#6	1.502
#7	2.044
#8	2.670
#9	3.400
#10	4.303
#11	5.313
#14	7.65
#18	13.60

Figure 3.1Ha

Bars are taken off by sizes and locations (such as footings, walls, columns, etc.). A waste, bending and overlapping factor, such as 15%, is added to the total weight. For mesh, 8 to 10% is added for overlapping.

For the examples given in this book, the unit costs in Figure 3.1Hb will be used. They include labor, material, equipment, and subcontractor's markup.

Typical Unit Prices for Reinforcing Steel in Place

Complexity	Light Bars #3 to #7 Per lb.	Heavy Bars #8 to #14 Per lb.	Mesh 6 x 6-10/10 Per sf
A Simple	.45	.35	.19
B Medium	.51	.40	.22
C Complex	.57	.47	.25

When hoisting is required, add .02/lb.

Figure 3.1Hb

Example:

Light bars	A	complexity	13,000 lb	@ .45 =	5,850
	C	"	9,000 lb	@ .57 =	5,130
Heavy bars	B	"	17,500 lb	@ .40 =	7,000
Mesh	A	"	30,000 lb	@ .19 =	5,700
					23,680
		Sub Markup @ 20%			4,736
		Total			28,416

Figure 3.1Hb is an aid in deciding, in a very general way, the degree of complexity. Costs may, be shaded between them.

Figure 3.1Hc

Complexity in Placing Rebar

A. Slabs on ground
Suspended plain slabs
Walls above grade-long, straight runs
Walls above grade-short, or curved

B. Foundation stem walls
Footings-continuous
Footings-spot column footings

C. Columns
Beams & girders
Pan-joist slabs
Steps & stairs
Small items with many cuts and bends
Very difficult access items

I.
Masonry

A budget for masonry work can be dependable if the estimator regularly practices and compares his budgets with competitive bids. To make a budget, the following information is necessary:

1. The kind of brick or block.
2. The size of brick or block.
3. The quantity.
4. The height of walls, columns, and pilasters.
5. Special inserts, control joints, etc.
6. Openings requiring lintels, sills, shoring, etc.
7. Waterproofing, integral color, face finishes.
8. Special banding.
9. Extent of solid grouting.
10. Reinforcing, type, size and spacing.
11. Special jointing.

Figure 3.1Ia lists a few typical masonry unit costs and is followed by an example of a masonry budget which is based upon this assumed cost data.

Unit Prices for Masonry Work, In Place					
Brick walls-Common brick		8″ (4″ face + 4″ backup)			11.96/sf
		12″ (4″ face + 8″ backup)			19.05/sf
Brick veneer-standard brick		(red face brick)			7.48/sf
glazed face					10.33/sf
Concrete block Sand aggregate	Thickness	Regular Block w/Reinf	Lite Wt Block w/Reinf	Add for Solid Grout	
Backup	4″	4.60/sf		.30/sf	
	6″	4.95/sf		.40/sf	
	8″	5.41/sf		.50/sf	
	12″	6.69/sf		.90/sf	
Foundation walls	6″	4.87/sf		.40/sf	
	8″	5.30/sf		.50/sf	
	12″	6.58/sf		.90/sf	
Walls	4″	4.64/sf	4.86/sf	.30/sf	
	6″	5.08/sf	5.24/sf	.40/sf	
	8″	5.56/sf	5.72/sf	.50/sf	
	12″	7.06/sf	7.22/sf	.90/sf	
For fluted block add		1.10/sf			
For average color add		.20/sf			
For highrise work add		.22/sf			
For tooled joints add		.15/sf			

Figure 3.1la

Example of a Masonry Budget

1. Regular (sand aggregate) concrete block foundation walls 8″ x 16″ x 8″, solid grouted, standard reinforcing, trowel joints   1,100 sf @ 5.80 =   6,380

2. Regular concrete block 8″ x 16″ x 8″ walls ungrouted, standard reinforcing, trowled joints   2,800 sf @ 5.56 =   15,568

3. Lightweight concrete block walls 8″ x 16″ x 6″, ungrouted, fluted face integral color, tool joints   900 sf @ 6.69 =   6,021

4. Common brick wall, 4″ thick backup and 4″ thick face brick   1,800 sf @ 11.96 =   21,528

5. Scaffold and hoist for portions of the above   2,000 sf @   .22 =   440

           49,937

Sub Markup 20%   9,987

Total   59,924

3.2 Dealing with subcontractors

In this book, estimating has been deliberately separated, as far as possible, from bidding. However, the estimator needs indications of the probable sub-bid coverage in order to plan his own estimating and budgeting work. Regarding each trade the following questions need answering:

1. Will there be one or more bids?
2. Will the bids be dependable (competitive)?
3. Will the bids cover the entire scope of the bidder's trades and sections of the specs, or will some of them be incomplete?

These questions cannot always be answered positively, but any indications are valuable guides to the estimator in selecting the trades, which will require budgeting.

While seeking information, the estimator frequently exchanges clarifications with subcontractors on such details as the following:

1. Furnish materials only?
2. Include sales taxes?
3. Unload trucks at jobsite?
4. Responsibility for storage and protection.
5. Include or exclude certain items.
6. Interpretations of drawings and specifications.
7. Include hoisting?
8. Include scaffolding?
9. Dates for beginning and ending of subtrade work.
10. Terms of payment and bonding.

In answering subcontractor's questions, the estimator, sets standards for the bidding.

Good rules for dealing with subs include:

1. Encourage subs to submit their bids on the same basis, so that they may be quickly and correctly compared.
2. Encourage subs to submit written proposals (flyers) prior to the bid deadline to accomplish four things:
   a. A thorough analysis of the proposed coverage.
   b. Opportunity beforehand for mutual discussion and/or alteration of terms of the proposed bid.
   c. Written proof of intention, to avoid possible later disputes.
   d. Save time at the crucial hour of bidding, by eliminating the need for detailed discussion.

From these inquiries and discussions with subs, the estimator is able to decide for which trades, or portions of trades, he should prepare budgets, and how much time and care is warranted in their preparation.

3.3 Dealing with material suppliers

The rules regarding subcontractors apply generally also to material suppliers, except that quotations by suppliers usually are needed and received earlier. No quotations from suppliers are binding upon them until the bid deadline arrives, nor is the estimator obliged to use any of the quotations. However, once the deadline is past, there is a certain obligation on the part of the supplier to stand behind his previous quotation. Up to the moment of bidding, both estimator and suppliers (within ethical boundries) are free to make changes, because the estimate is in a "plastic" state (see Section 1.12—Ethical Considerations).

Quotations given without qualifications generally are assumed as firm, and unless the supplier changes them before the bid deadline, may be depended upon, particularly if they agree closely with other suppliers' quotations.

Good rules for dealing with material suppliers include:

1. Request quotations for every project because of changing prices and varied delivery conditions.
2. Whenever possible, request quotations for guaranteed periods, preferably for the duration of the project.
3. Keep all quotations confidential.
4. Avoid doing business with only one supplier.
5. Maintain goodwill.
6. Alert a supplier if his quotation is glaringly out of line with those

of other suppliers but at the same time keep confidential the amounts of the other quotations.

3.4 Plug-in figures and allowances

Plug-in figures serve as temporary fillers to complete an estimate. They are used in one or more of the following conditions:

1. When sub-bids are expected, but have not yet been received. Sometimes a subcontractor will propose a plug-in figure as a courtesy and convenience to the general contractor.
2. When the estimator has no time to make a budget.
3. When the estimator lacks sufficient experience to make a budget.
4. When details and/or specs are nonexistent, or insufficient for the making of a budget.
5. When conditions are unknown.
6. When there is no other way to complete an estimate.

Before the bid deadline, every possible effort is made to replace plug-in figures with dependable estimates, budgets or sub-bids.

Figure 3.4a is an example of comparative estimate summaries. Plug-in figures are in parentheses; in an actual estimate they may be in red pencil, as a reminder of their undependable status.

Trade	Working Estimate		Final Estimate	
	Estimate	Status	Estimate	Status
Demolition	16,000	budget	17,700	sub-bid
Earthwork	145,000	budget	128,000	sub-bid
A.C. Paving	92,225	sub-bid	92,225	sub-bid
Fencing	(5,000)	(plug-in)	6,350	sub-bid
Site concrete	22,000	budget	22,000	budget
Struc concrete	475,155	estimate	475,155	estimate
Reinf steel	82,170	sub-bid	82,170	sub-bid
Misc metal	(33,000)	(plug-In)	27,500	sub-bid
Carpentry	28,315	estimate	28,000	estimate
Sheet metal	7,300	sub-bid	7,300	sub-bid
Roofing	19,500	sub-bid	19,500	sub-bid
Doors & frames	13,000	budget	14,960	sub-bid
Acoustic tile	8,400	sub-bid	8,400	sub-bid
Resilient flr'g	11,650	sub-bid	11,650	sub-bid
Painting	(30,000)	(plug-in)	31,375	sub-bid
Mechanical	275,000	sub-bid	275,000	sub-bid
Electrical	184,350	sub-bid	184,350	sub-bid
	1,448,065		1,431,950	

Figure 3.4a

Comparison of the working and final estimates in Figure 3.4a shows the following changes in status (dependability of the figures):

	Working Estimate	Final Estimate
Sub-bids	8	14
Estimates	2	2
Budgets	4	1
Plug-in figures	3	0
	17	17

In the final estimate, all plug-in figures are replaced with firm sub-bids; only one budget (site concrete) remains, but this work lies within the expertise of the contractor, as do the two estimated quantities. Consequently, the final estimate is made as "safe" as possible except for consideration of contingency allowances which are taken up in the next section.

The final estimate can hardly be called an "estimate", because it has metamorphosed into a compiled list of sub-bids. Except for the remaining two estimates and one budget, all of the estimator's work has disappeared. It is possible for the bidding stage to completely wipe out *all* of the estimator's work, and the contractor would be what is called a "broker". This seldom happens in actual practice, and there may be legal restrictions to discourage it; but even when it happens, the work of the estimator leading up to that conclusion is essential.

3.5 Contingency allowances

The estimator predicates the cost of a project upon the experienced "normal" course of events. But experience also shows that departures from the normal can occur, and they may be partly predictable, or completely unpredictable. The estimator builds into his figures allowances for the partly predictable, such as extremes of weather and pending labor strikes (see Section 2.13—Analyzing a Project for Desirability); but what of the totally unpredictable, such as the following examples?

1. Underground water, rock or buried concrete.
2. Wildcat strikes.
3. Arbitrary material price increases.
4. Drawings which appear correct, but which prove in the field to be uncoordinated.
5. Floods, earthquakes, or fires.
6. Structural collapses.
7. Bankruptcy or default of subcontractor.
8. Loss of an essential technician or supervisor.
9. Faulty layout in the field and subsequent cost to repair or replace.
10. Installation of noncomplying material, and cost of replacing.
11. Misinterpretation of plans and specs.
12. Mistakes in estimating.
13. Failure or inability of manufacturer to supply critical material or equipment on schedule.
14. Numerous change orders and inefficient expediting by owner/representative.
15. Theft and/or vandalism.
16. Litigation.

Allowances for such unpredictables as the above include many strategies which are not the immediate business of the estimator, but which he needs to know to complete the estimate. Among them are:

1. Subletting a trade and thus transferring the "gamble" to another party.
2. Requiring a performance bond of a subcontractor, thus ensuring against default.
3. If possible, qualifying the bid at bid time (not always permitted).
4. Attempting to obtain from the A & E clarifications and/or addenda.
5. Obtaining "special risk" insurance.
6. Employing engineering consultants to minimize field errors and slip ups.
7. Making budgets of costs of questionable subtrades in order to

assess the "gamble" (see Section 3.1—Budgeting Subtrades).

8. Including in the estimate an actual sum as an allowance for contingencies.

Some of the above strategies are the same as allowances. The estimator inserts them in specific items, such as the general conditions. Also, a subtrade bid which is higher than the estimator's budget is a hidden allowance for contingency. That is also true of subcontractor performance bonds, consulting fees and non-required insurance.

All of such hidden allowances should be taken into account if an actual contingency allowance is to be used. In the practice of competitive bidding, such an outright allowance is rarely used; but in preparing the estimate, it is good practice for the estimator to propose a "realistic" contingency figure to use in the event that one is warranted at bid deadline.

The unpredictable cannot be estimated. At best it is only a judgment—an educated guess (see Section 1.8—Judgment), but here are some rules of thumb:

1. Isolate the items of probable risk, and their probable cost.
2. Subtract the amounts of hidden allowances.
3. Compare to total markup amount.
4. Use as a contingency allowance that amount which exceeds half of the markup.

Example:

Total cost of a project, 5,000,000; markup 250,000

Amount left on table by "risky" subs	150,000
Cost of overhead due to completion time needed in excess of that required in specs	15,000
10% of contractor's own work (not sublet) allowed to cover unpredictable contingencies	50,000
Subtotal	215,000
Less hidden allowances within the estimate	30,000
Hypothetical risk	185,000
Less ½ of markup	125,000
Proposed contingency allowance	60,000

This approach to a contingency allowance is a rationalized compromise between the various risks, hidden allowances and the markup (which may be thought of loosely as an indirect contingency allowance). Such final allowance may be incorporated in the estimate in one of two ways: (1) as a specific item, or (2) an addition to the markup, thus:

In the example above the markup of 250,000 is increased by 60,000 to 310,000 and the contingency allowance, as such, is simply forgotten. A better way yet is a combination of the two—a specific item placed so near the markup as to be in effect part of the markup, thus:

Estimated cost	5,000,000
Markup	250,000
Contingency allowance	60,000
Total Estimate & Bid	5,310,000

In this position, the allowance can easily be revised, or eliminated, without the bother of changing any other figures.

Competition influences the amount of a contingency allowance and tends to prevent full protection from risk. The rationalization is this: chance works both ways. The concern is with loss; but chance can also produce increased profit; realization of this possibility tends to reduce a contingency allowance, or obliterate it, or even convert it to a "cut".

A cut is a reverse contingency allowance, based upon the prediction of *good* fortune. If in the above example, the following conditions are judged to exist:

Risk in subcontract bids	none
Slack in certain items	45,000
Hidden allowances	30,000
Savings from expectations of early finish	15,000
Hypothetical amount overestimated	90,000
Less 10% of contractor's work retained for contingency	50,000
Proposed cut	40,000

The total bid would be: 5,000,000 + 250,000 - 40,000 =   5,210,000

3.6 Presentation meetings (hashing over)

A few days before the bid deadline, the estimator reviews his work in detail with the company's management staff in order to:

1. Furnish them the opportunity to accept or change the figures.
2. Improve the estimate by inviting other's ideas and constructive criticism.
3. Discover hidden errors and conditions which the estimator might not have known or thought of.
4. Involve others in the project and start the teamwork which will carry on from this point to the bid deadline and beyond.
5. In effect, deliver the tentatively finished work (the estimate) to the company. It is a fair assumption that cost items, which are not challenged or changed, are tacitly accepted by all.

While explaining all parts of the estimate to the people who are seeing it for the first time, the estimator must also describe the project, explain the drawings, and emphasize the "trouble areas" (this routine is not unlike a military briefing).

In preparation for the meeting, all important points and items are "flagged" with red pencil. These marks are erased as each item is settled by discussion and agreement or decision. All items not settled are left flagged until they can be resolved prior to the bid deadline, or by incorporation in a contingency allowance.

Of the five estimating forms shown in Section 1.10—Forms, Formats and Systems of Estimating, number 2 (miscellaneous notes) and number 5 (worksheet) are made with the presentation meeting in mind. Their use isolates the trouble areas and displays in detail the estimator's proposed solutions.

At the start of the meeting the estimate is only a proposal; at the meeting's end the estimate is an accepted fact (except for any remaining red-flagged items). Guidelines for the estimator in conducting the presentation meeting include:

1. Drawings, specs and estimate sheets should be marked and flagged with colors to draw attention to important items of cost, or potential trouble.
2. Follow an order of logic and clarity, giving priority to those items

most important to the contractor. A typical order is suggested as follows:

    a. Demolition and alteration work.
    b. Structural concrete and associated excavation work.
    c. Site concrete work.
    d. Carpentry and millwork.
    e. Miscellaneous trades which are optional as to subletting.
    f. Trades which are budgeted for cost.
    g. Analyze subtrades, the points of separation between them, and items of possible exclusion by sub-bidders.
    h. Special attention to site conditions and earthmoving methods.
    i. General conditions and their costs.
    j. Bid items (alternate bids) and addenda.
    k. Bidding conditions, competition, time for completion, markup, contingency allowance, etc.
    l. Distribution among those present of bidding duties, further research and pursuit of firm quotations.

3. Draw upon the field experience of those present at the meeting.
4. Avoid "selling" or defending the previously estimated figures; readily change any of them in the light of discussion.
5. Welcome the most ruthless cross-examination. Recognize the meeting as a period of criticism, and strive to be the most critical person present.
6. Introduce as many *other* sources of cost information as possible, such as past job cost records, previously used unit prices, cost reference book data, expert advice, catalog data, etc., to support the estimate and prevent sole reliance on the estimator's personal cost values.
7. For items not confirmed by *other* cost data, show back-up reasonings and calculations (sketches, if needed for clarification) on the worksheet.
8. For certain items, make a simple list of the estimated costs used in previous projects for comparison and display.
9. Offer extensive details of proposed methods of doing work items, such as choice of equipment, choice in formwork design, shoring, scaffolding, conveying, hoisting, and so forth (see Section 1.14—Selecting Methods of Construction).

The presentation meeting is a period of transition from the estimating to the bidding stage. The relatively impersonal work on drawings and specifications gives way to increasing work with people. The more clearly and thoroughly the estimator transmits the information, the more successfully will the aims of the meeting be fulfilled.

3.7 Firming up the estimate

At the presentation meeting, parts of the estimate are "accepted", and other parts (the red-flagged items) are yet to be revised by recomputing, by researching, by replacing with firm quotations, or by completely re-estimating using a different method of construction than previously assumed. Until these final revisions are made, the unsuitable figures act as plug-in-figures (see Section 3.4—Plug-In Figures and Allowances).

When all changes have been made, and the red flags obliterated, a final checking of all extensions (multiplication, division, addition, subtraction) is done, preferably by some person other than the original estimator.

Except for addendum changes and improved material cost quotations, the estimate is now completed. There remains only the bidding procedure.

3.8 Addenda

When the designer (A & E) finds it desirable, advisable, or necessary to make a change to the drawings and/or specs after they have already been issued to bidders, he issues an addendum. Since addenda are usually last minute changes, they require revisions to figures already incorporated in the estimate. The possibility of forthcoming addenda is one of the reasons for breaking down the costs into trades, categories, items and elements in relatively fine detail, as shown in Section 1.7—Divisions of the Estimate. By means of these breakdowns, small or large portions may be simply and quickly revised or eliminated.

Some useful rules regarding addenda are:

1. Large cost changes can result from a few words or numbers; therefore, the estimator should take the time to carefully read and compare addenda to the original drawings and specs.
2. Items or elements in the estimate, changed by addenda, should not be erased; they should be crossed out without losing their legibility and revised so that both old and new figures are shown for the record.
3. Changes should be made on the drawings and specs with colored ink so that both the original and the revised are compared and permanently recorded for the later convenience of the project super and crews.
4. Notify subcontractors whose work is affected by addenda as soon as possible.

3.9 Bidding the job

Often the estimator assumes charge of the bidding operation; but in his strict role as estimator, he holds an objective and disinterested view of the outcome of the bidding (see Section 1.6—The Estimator's Role in the Company and Section 1.6—Considering Competition). Since this book minimizes the subject of bidding, only a brief outline follows.

The time to do a proper job of estimating becomes more critical as the bid deadline approaches; therefore, the bidding system is designed for speed, simplicity, safeguards against mistakes and minimum motion in the receiving, analyzing, comparing, selecting of sub-bids, summing up and submitting (tendering) the bid.

The bidding is accomplished by a team of five or six people. The estimator's most important task is the analyzing and comparing of sub-bids.

The problems of bidding are common to all estimators, and although each develops his own techniques, the following objectives are shared:

1. Accomplish everything as early as possible in order to minimize haste in the last minutes before the deadline.
2. Utilize the rooms, furniture, equipment and personnel in the most efficient manner possible (see Section 1.18—Estimator's Working Conditions).
3. Use a format for tabulating, displaying, comparing and recording all sub-bids, trades and cost categories. It should be a one-sheet affair, as large as necessary, available to the eyes of all members of the bidding team.
4. Analyze and compare sub-bids, using previously prepared budgets

and plug-in figures to complete any portions excluded by sub-bidders. See Figure 3.9a for an example.

5. Sum up all figures and check them for accuracy well before the bid deadline to avoid mistakes caused by last minute haste.
6. To the total cost add contractor's markup (see Section 3.10— Overhead, Profit and Bond).
7. To the first subtotal add or subtract any last minute changes. See Figure 3.9b.

Sub-bids are often incomplete, faulty, or non-existent. An itemized budget may be substituted in part or entirely, as in Figure 3.9a.

Comparison of Sub-bids and Budget any Trade

Item	Budget	Sub A	Sub B	Sub C	Sub D
		101,000	115,000	130,000	97,000
1	6,500				*6,500
2	23,000				
3	18,000	*18,000			*18,000
4	5,525	*5,525			
5	2,700				
6	62,000				
7	9,275		*9,275		
Total	127,000	124,525	124,275	130,000	122,000

*Items not included by sub-bidders are filled in with values budgeted by the estimator.

Sub C is the least risk, but the highest bid.
Sub B is the second least risk and the second lowest bid.
Sub A is the third least risk and third lowest bid.
Sub D is the fourth least risk and the lowest bid.
Budget amount is the greatest risk.
Choice is between Subs B and D.

Figure 3.9a

In the last few minutes before the bid deadline, changes may be advisable. They are made most quickly and with the least chance of error by reduction to one simple additive or deductive amount, as follows:

Amount of bid before any final
adjustments, using the example in
Section 3.5 5,310,000

	Cut	Add
Structural steel		7,000
Electrical	15,000	
Earthmoving	13,000	
Painting		4,000
Subtotal	28,000	11,000
	Net Cut	17,000
Final bid amount		5,293,000

Figure 3.9b

3.10
Overhead, profit and bond

Overhead is the contractor's cost to operate his overall business. Since it is not a true cost item in any particular project, it does not appear in the estimate except as a part of the contractor's markup. As shown below, it is only roughly computable.

Profit is the contractor's motivation, without which there would be no bidding. It is also a necessity to the stimulation of business growth, without which a construction business cannot survive. Only profit can offset the ongoing expense of overhead. Unlike overhead, profit is not computable; but there are some guidelines, as suggested below, for basing the amount in a bid.

Bond cost, unlike overhead and profit, is entirely computable. Since the bond cost is based upon the total amount of the bid (contract), it is the last figure in the estimate.

The formula differs with contractors and bonding companies, but a typical one, for use in the examples given in this book, is shown in Figure 3.10c.

Since these three items of markup are developed in the final moments of the bidding they are not, strictly speaking, a function of estimating, and therefore will be only briefly explained here. In practice, overhead and profit is usually added to the estimated cost by a single, lump sum amount. Given the fact of competition, profit is limited to a "reasonable" amount. A table, such as Figure 3.10b, may be constructed to serve as a guideline for appropriate markup (not including bond cost).

The variable factors which affect the desirability of a project (see Section 2.13—Analyzing a Project for Desirability) contribute to the profit portion of the markup, while the overhead portion remains constant. Therefore, before studying Figure 3.10a, guidelines for computing overhead and prorating it into individual projects should be considered.

Assume a construction company limited to bonding and organizational capacity of approximately 15,000,000 gross, average complexity, construction work annually, and an overhead of 300,000. If the company could be assured of obtaining that full volume of work, the amount to be included in each bid for overhead would be 300,000/15,000,000 = .02 or 2% of estimated cost. However, maximum procurement is uncertain, so assume half and use:

$$300,000/7,500,000 = .04 \text{ or } 4\%.$$

Every project, regardless of its desirability rating, would include the apportioned overhead; in this case 4% of the total estimated cost. Figure 3.10b includes 4% as a constant in all the percentages shown.

Profit, unlike overhead is strongly influenced by the desirability rating of the project. Figure 3.10b is constructed of two main influences: the desirability rating and the size in dollars of the project. There are several reasons for the size of the project to rate a special influence on markup; the main one is that competition forces a diminishing ratio between size of project and amount of markup.

A properly estimated project should not depend upon any portion of the profit for protection against risk, although it may affect the amount of a contingency allowance (see Section 3.5—Contingency Allowance).

Experience in bidding indicates a range for profit which is normally permitted by the market. Using the previous example of 4% for overhead, the minimum markup should be something greater than 4%, if it is to include profit; the maximum may be as much as 25%, depending upon the desirability, size and nature of the project.

Using a table similar to the following, Figure 3.10a, the desirability rating of a project may be judged.

Project Desirability Rating Scale

Condition	Rating 0 to 100
1. Location: distance from home office to project (the nearer, the higher the rating)	_____
2. Character of the project (complexity/simplicity)	_____
3. Quality of drawings & specs	_____
4. Quality of available supervising personnel	_____
5. Quality of available workmen	_____
6. Co-operativeness of A & E, owner, inspector	_____
7. Length of project: completion time (the shorter, the higher the rating)	_____
8. Proportion subbed out (the more, the higher the rating)	_____
9. Company's need for work	_____
10. Economic conditions (prospect for availability of work in the future)	_____
Subtotal	_____
Divide by 10	_____
Total rating	_____

Figure 3.10a

Any rating over 90, or under 20 should not normally be used but should be reserved for unusual conditions in order to provide special influence on the total rating. In the following example the company's need for work is unusually great, so the full impact of 100 is used.

Example:

A project is approximately 7,000,000 in size and the desirability rating is as follows:

1.	The project is 200 miles distant from office	30
2.	The project is very simple to construct	90
3.	The drawings & specs are of average quality	75
4.	Known excellent superintendent is available	90
5.	Workmen available are above average in abilities	85
6.	Cooperativeness of A & E, owner, etc., good	85
7.	Completion time, slightly long	70
8.	Large proportion of the work to be sublet	90
9.	Need for work is unusally great	100
10.	Economic outlook is mediocre	85
		800

$$\frac{800}{10} = 80$$

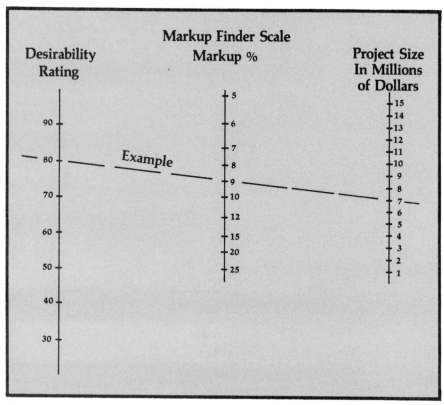

Markup Finder Scale

Desirability Rating	Markup %	Project Size In Millions of Dollars

Figure 3.10b

On Figure 3.10b draw a line connecting project size of 7,000,000 and desirability rating of 80. On center scale find 9% markup. If, in this example, the need for work had been 0, the desirability rating would have been 70, and the markup would have been 10%. The total difference in dollars would have been 70,000. This may seem small compensation for such an extreme difference in need, but a rating of 70 is low, its resulting 10% markup is high, and competition is unlikely to permit the luxury of full proportionate compensation. Another point is this: When the need for work is 0, it means that work on hand is enough to provide the entire annual overhead plus a reasonable profit. The 4% included in the markup for overhead could be eliminated, reducing the markup to only 6%. The paradox is that the less work is needed, the lower (to a point) the required markup but the increased independence of the bidder causes him to ignore that fact.

The conclusion is that markups, found by use of such a scale as Figure 3.10b, require further adjusting. They are idealistic, but they provide the initial range. Even without such a scale for reference an experienced estimator's mind, working computer-fashion, finds the initial range and makes appropriate adjustments. The purpose of this exposition is to provide a rational foundation for the markup, rather than what may appear as an arbitrary decision.

Finally, to show how overhead, profit and bond are applied to a typical estimated cost, the following example is given:

Estimated cost of a project	5,222,200
Overhead & profit (desirability factor 85%) from Figure 3.10b = 8.5%	443,887
Subtotal	5,666,087
Bond cost from Figure 3.10c = .005 = .5%	28,330
Total Amount of Bid	5,694,417

Formula for Computing Cost of Bond		
Estimated Cost of Project		**Multiply By**
From	**To**	
0	1,000,000	.0075
1,000,000	1,500,000	.0063
1,500,000	2,000,000	.006
2,000,000	2,500,000	.0058
2,500,000	3,000,000	.0056
3,000,000	4,000,000	.0055
4,000,000	5,000,000	.0052
5,000,000	6,000,000	.005
6,000,000	7,000,000	.0048
7,000,000	9,000,000	.0047
9,000,000	11,000,000	.0046
11,000,000	and over	.0045

Figure 3.10c

3.11 Follow-up

After the bidding, certain procedures are required of the estimator. He is responsible for the saving and filing of the estimate sheets, drawings, specs and all revisions.

For successful bids (contracts obtained) the estimator has certain immediate duties transitional to the physical activities of the construction work. He must communicate and pass along to project superintendents, subcontractors, material suppliers and others, information and instructions with which he, more than any other person, is familiar.

These duties include:

1. Write (rough draft) subcontracts and purchase orders.
2. Compile schedules of prices (basis for monthly payments for work performed).
3. Compile cost-record sheets (for feedback cost information).
4. Design construction progress schedules (sometimes this is sublet to engineer specialists).

Not so immediate, and continuing throughout the construction period, are:

1. Computing change orders.
2. Analyzing cost records.
3. Miscellaneous followup.
    a. Submitting for approval shop drawings, samples, etc.
    b. General expediting.
    c. Jobsite visitations.

All of these duties have little to do directly with estimating, and therefore discussion of them will be brief.

Writing subcontracts and purchase orders by the estimator is limited to the technical details, while the more legal are left to others in the company. Guidelines for the estimator include:

1. Construct a check list classifying all trades, categories, items and elements in the project under the following headings:
    a. Subcontract
    b. Purchase order
    c. To be provided by superintendent
    d. To be provided by office

e. To be provided by others (items not immediately assigned, and awaiting decision as to assignment)

2. Identify the scope of work, as to general trade, sections of the specs, drawing sheets and detail numbers, when clarity requires.
3. Describe for inclusion in subcontracts any items which are potential subjects for disputes, such as:
   a. Cleanup and disposal of debris.
   b. Unloading, storing and protecting of materials.
   c. Hoisting and scaffolding.
   d. Temporary power, and water service.
   e. Laying out of work.
   f. Caulking and sealing.
   g. Acknowledge materials which are to be delivered only.
   h. State whether tax is included or not.
   i. Cutting and patching work.
4. Assign and describe items which customs-of-the-trades do not clearly delegate (see Section 1.21—Customs-of-the-Trades), such as:
   a. Quality of the materials and workmanship.
   b. Points of departure between subtrades (for instance, between sheet metal and other steel trades).
   c. Concrete work for mechanical, electrical and other trades.
   d. Saw cutting and patching of pavement.
5. Specify time limitations when important.
   a. For submittals of samples and shop drawings.
   b. For delivery of materials to jobsite.
   c. For performing particular phases of the work.
   d. For cooperation with the overall project time schedule.
6. Assign responsibility for special inspections and testing of materials.
7. Request special insurance when warranted.
8. State the amount (dollars) and terms of payment.
9. Acknowledge any items stated by the subcontractor as qualifications to his bid.

Subcontracts and purchase orders should be written with the realization that persons other than the original bidders may later interpret and respond to them. The need for thoroughness and accuracy lies in the fact that written subcontracts supercede all previous verbal quotations and agreements.

Compiling schedules of prices for periodic payments of completed work is a standardized procedure. It is customarily done in the format of the A & E or owner.

Compiling cost record sheets is a matter of using the company's preferred method.

Designing of construction progress schedules is not always the estimator's duty, but no one is better suited for the task than the estimator. The format is sometimes defined in the specs, but whatever the form, the estimator already has a start in the work he has under Section 2.19—Construction Progress Schedules. The outline started there, may be used as a guide to the construction of the final, fully detailed schedule.

Computing change orders follows the same format as the original estimate, with breakdowns for labor, material, equipment, subcontract work, overhead, profit and bond.

Analyzing cost records-see Section 3.12—Feedback.

Miscellaneous follow-up, carried out by the estimator as time permits, includes such duties as submitting shop drawings and material samples to the A & E for approval, general expediting, and project visitations for purposes of inspection.

3.12 Feedback

Information of the actual costs, fed back to the estimator from field supervisory personnel may be divided into two types, namely (1) cost record data, and (2) general information regarding any unforeseen difficulties, the actual methods and equipment used, and so forth. Cost records have more value when they are accompanied by information of special conditions affecting them.

Any cost record is better than none. The estimator is eager for any scrap of useful information; but the value increases as more time and carefulness are invested in them by all who participate. In the field, cost records take second place to the super's more essential work of managing the workmen and subs, and as a rule their quality is proportionate to the interest displayed by the estimator.

A cost record system may be designed for any degree of simplicity or complexity desired. At its simplest, it is only a spot-checking of selected items and does not serve to control the actual cost of the construction work; at its most complex, which will hereafter be called "the complete method", it includes *all* of the cost items in the project, no matter how trivial, and indicates their trends toward running over or under the estimated costs. Prompt reporting permits special actions to control the costs while there is still time.

Figure 1.9a shows the simple method, which gives only the final recorded costs, too late to apply controls, but still useful for reference in the future estimating of similar items.

Figure 3.12a shows a portion of a complete cost record, which serves the company's need for cost control. Because these costs are achieved under the pressure of close supervision, they should be regarded as *minimum*, while those in Figure 1.9a are average.

In Figure 3.12a, general information notes, if volunteered by the super, would further explain the differences between actual and estimated costs (why they occured).

To save time, this system may be computerized; however, the longhand method has the advantage of close, personal, indepth study.

When a project is completed, to simplify the reference, the final cost record sheets may be placed in a book, or file folder with the cost records of other projects of similar type. Some examples are:

> Warehouses
> Shopping centers
> Industrial buildings
> Medical, dental, hospital buildings
> Schools and universities
> High rise office buildings
> Dormitories & barracks
> Restaurants & mess halls
> Waterfront & pier work
> Water & sewer treatment plants
> Site preparation work

So compiled, these records may be indexed so that specific trades may be referred to in a minimum of research time.

Example
Portion of a Cost Record

		Estimated					Actual Cost			
	Q	l	m/e	L	M/E	Q to date	L	M/E	l	m/e
1. Pile Caps										
a. Layout	109 ea	9.25	3.20	1,008	349	109 ea	920	305	8.44	2.80
b. Mach exc	1,010 cy	—	4.75	—	4,798	996 cy	—	5,210	—	5.23
c. Hand exc	109 hrs	15.10	—	1,646	—	120 hrs	1,820	—	15.17	.37
d. Forming	7,820 sf	2.20	.37	17,204	2,893	6,000 sf	11,800	2,220	1.97	.37
e. Concrete	483 cy	7.60	58.00	3,671	28,014	372 cy	2,678	21,276	7.20	57.19
f. Backfill	550 cy	5.21	2.29	2,866	1,260	0	—	—	—	—
g. Disposal	460 cy	.80	1.60	368	736	450 cy	200	820	.44	1.82
2. Slab on Grade										
a. Fine grade	20,220 sf	.07	—	1,415	—	10,000 sf	750	—	.075	—
b. Set screeds	20,220 sf	.14	.05	2,831	1,011	12,000 sf	1,620	600	.135	.05
c. Concrete	395 cy	8.00	58.00	3,160	22,910	190 cy	1,420	10,920	7.47	57.47
d. Finish/cure	20,220 sf	.26	.03	5,257	607	9,000 sf	2,250	370	.25	.04
3. Walls										
a. Formwork	6,245 sf	2.90	.80	18,111	4,996	3,000 sf	9,660	2,730	3.22	.91
b. Concrete	170 cy	9.40	60.00	1,598	10,200	65 cy	715	3,874	11.00	59.60
c. Hoisting	170 cy	—	11.50	—	1,955	65 cy	—	705	—	10.84
d. Point/patch	6,245 sf	.12	.02	749	125	2,200 sf	220	68	.10	.03
e. Rub/grind	3,000 sf	.60	.09	1,800	270	800 sf	600	120	.75	.15

Figure 3.12a

Analysis of Figure 3.12a reveals:

Work Item	Completion	Over/Under
Pile caps-layout	completed	underestimate
Machine excavation	completed	overestimate
Hand excavation	completed	overestimate
Formwork	80%	underestimate
Concrete work	75%	underestimate
Backfilling	none	
Disposal of excess dirt	completed	underestimate
Slab on grd-fine grading	50%	running over
Setting screeds	60%	running under
Concrete work	50%	running under
Finishing & curing	45%	on estimate
Walls-Formwork	50%	running over
Concrete work	33%	over on labor
Hoisting	33%	running under
Pointing & patching	33%	running under
Rubbing & grinding	25%	running over

Part 4 Estimating A Typical Project

Let us imagine that a construction project has just been released to interested bidders, and (since complete plans and specs cannot be included in a book such as this) is briefly sketched and described hereafter.

Project: Aircraft Parts Storage Warehouse, a one-story steel framed and tilt-up concrete building with attached concrete masonry office wing, bordered on two sides by fenced-in paved parking area and on the other two sides by landscaping; see Figure 4a.

This theoretical project provides a nice combination of engineering and building construction, both heavy and light, involving many architectural trades. Material quantities and simple drawing details will be given as they occur in the order of the estimating.

The symbol (*) is used for items and unit prices for which no reference is given in this book, and for which outside sources, such as direct quotations, reference books, field cost records, etc., provide information. Since these cases occur frequently in all estimates, some are included in this example for the sake of realism.

Step 1 Analyze the project for desirability

Section 2.13

This first scan produces only a general impression, but sufficient to provide guidelines for a method of estimating and an appropriate attitude toward the project. It sets a tone for all subsequent estimating, for tightness or looseness in the figuring.

In this project, let us assume that the questions, answers, and rough analysis goes as follows:

Q. Where is the project located?
A. 90 miles distant from the home office (negative point).
Q. What is the size (dollar value)?
A. The advertised cost range is 5,000,000 to 7,000,000 dollars. A quick impression would place its value near the lower figure.
Q. What is the required completion time?
A. 600 Calendar days, or approximately 20 months.
Q. What are the liquidated damages?
A. 750 dollars for each day the project completion is delayed beyond 600 days.
Q. Is the completion time too short? Are the liquidated damages too great?

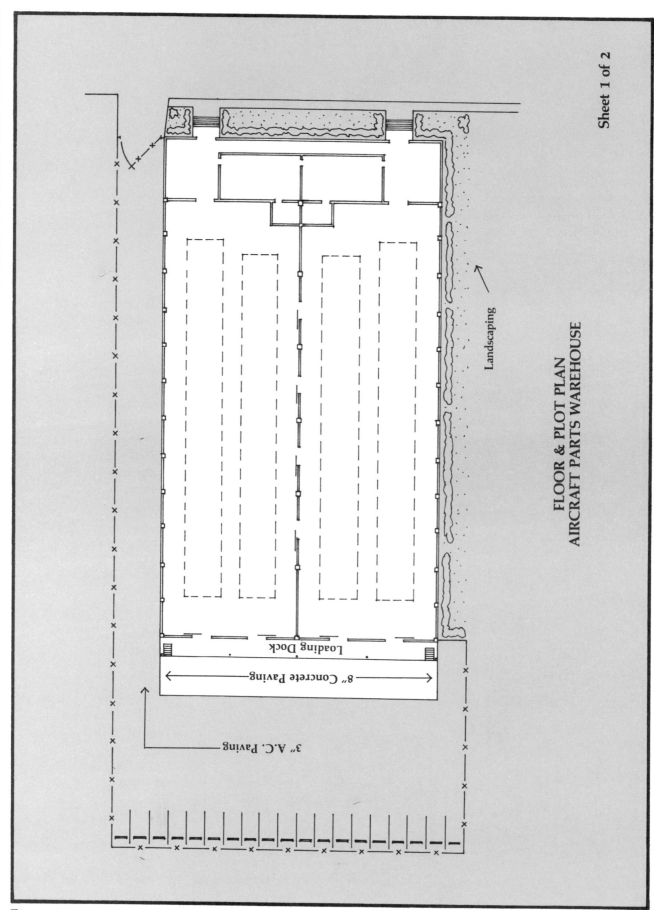

FLOOR & PLOT PLAN
AIRCRAFT PARTS WAREHOUSE

Landscaping

Loading Dock

8" Concrete Paving

3" A.C. Paving

Figure 4.1a

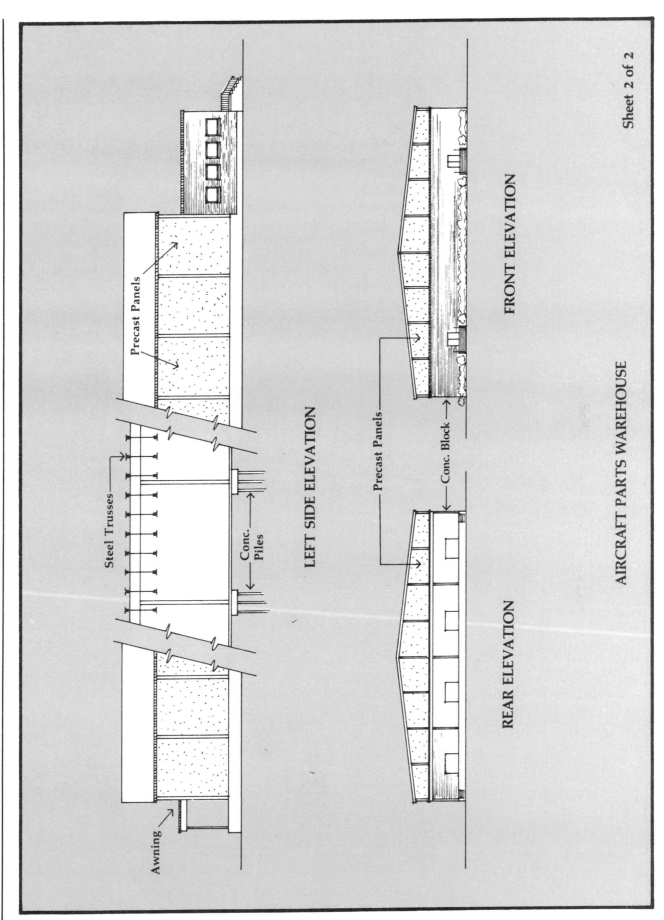

Figure 4.1a (cont.)

A. Off-hand impression is negative to both parts of the questions. They both appear reasonable.

Q. Is the size desirable?

A. Yes. It is ideal for the company's present need.

Q. Is the type of construction desirable?

A. The company is experienced in this type of construction, and has available a competent superintendent.

Q. What is the quality of the drawings and specs?

A. On first appearance, this is hard to judge. Neat draftsmanship does not guarantee good planning nor untidy draftsmanship bad planning. Let us say that a mediocre quality is indicated.

Q. What is the company's need for work?

A. The present need is not great, but economic forecasts indicate a slowing down of construction activity.

Q. In view of the above, and other considerations as listed in Figure 3.10a, without at this time attempting any itemizing of values, what is the impression expressed on a scale of 0 to 100 of the project's desirability?

Figure 3.10a

A. Well above average—say, 88. We will keep this in mind and be prepared to modify it if later information advises. All of this will be considered in greater detail when the markup is decided.

Section 3.10

Step 2
Scheduling the time to estimate

Section 2.14

Figure 2.14d

We are able to tell from the scope of work of each trade, as shown in the drawings and specs and with the following bar graph the approximate time required for one estimator to take off, price out and follow through to the hour of bidding. If two estimators were to work as a team the number of days would be reduced.

The project is expected to cost about 5,500,000 dollars, and if it is judged at "A" complexity on Figure 2.14d, the *net* hours to estimate would be about 127. If 5 net hours of estimating work are possible daily, then 127/5 = 25.4 days divided as follows:

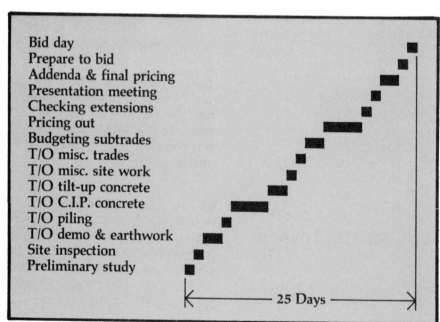

Bid day
Prepare to bid
Addenda & final pricing
Presentation meeting
Checking extensions
Pricing out
Budgeting subtrades
T/O misc. trades
T/O misc. site work
T/O tilt-up concrete
T/O C.I.P. concrete
T/O piling
T/O demo & earthwork
Site inspection
Preliminary study

|← —————— 25 Days —————— →|

Figure 4.2a

Now that we have judged the project as desirable to bid, and the time to do the estimating work is approximately 25 days, we must fit it, if possible, into the company's current estimating work schedule.

Figure 2.14b

Projects A, B and C are underway and the black squares represent their bid dates. Our contemplated project, (dash lines), cannot be started until project A is completed, but then only 18 days remain. The 7-day shortage may be made up by shifting the estimator from project B, at its completion, to assist with the estimating work.

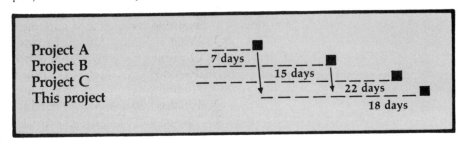

Figure 4.2b

Step 3 Analyze the bid schedule

Section 2.15

The bid schedule, as found in the specs, requests:

1. A Base Bid consisting of a lump sum price for all of the work except the following three additive alternate bids.
2. An Alternate Bid A, to furnish and install pallet storage racks.
3. An Alternate Bid B, to furnish and install three adjustable loading ramps.
4. An Alternate Bid C, to construct a section of concrete paving adjacent to the loading dock in lieu of A.C. paving.

For segregation purposes we note that:

Alternate Bid A consists of:
  a. subcontractor - add pallet storage racks.
Alternate Bid B consists of:
  a. subcontractor - add adjustable loading ramps.
  b. *Estimate by us - deduct* sand fill in and temporary slabs over pits which are to be constructed in the base bid.
  c. subcontractor - add electrical hookups to ramps.
Alternate Bid C consists of:
  a. subcontractor - *deduct* A.C. paving. Add excavation.
  b. *estimate by us* - add concrete paving.

Items to be estimated by us (not sublet) are in italics to emphasize that they are our foremost points of interest. Other italicized points of interest are certain items which are deductive. These should be printed boldly in red pencil, here, and in all future references to them. Those items which are subcontractor work may be left to the various competing subcontractors to note and quote; but, we need to have them clearly in mind in order later to deal with the confusion which is common on bid day.

Section 3.9

Step 4 Inspect the jobsite

Figure 2.16a

Since this project is reported to be 90 miles distant from our office, we can plan on a minimum of five hours for the round-trip inspection visit. In order to do a thorough job, and avoid having to make a second trip, we should study the drawings carefully beforehand, make notes for our guidance and prepare a site investigation check sheet.

Some of the conditions we look for, are these:

- Check the actual mileage for travel to/from the jobsite.
- Accessibility for equipment to the jobsite (will access roads be required?).
- Answer all the items on the site investigation check sheet.

- Check to see if all items requiring demolition are actually shown on the drawings.
- Obtain information such as dimensions, sizes and number of trees, thickness and condition of pavement, not shown on the drawings.
- Roughly plan the locations of construction yard, job office, equipment, etc.
- Look for clues to underground water, rock, etc.

Site Investigation Check Sheet

Project __Aircraft Parts Warehouse__   Bid Date __July 7, 1982__

Location __Boulder, Ca.__   A & E __G.H. & Tech Inc.__

1. Distance from home office __Checked by auto - 92 miles__

2. Subsistence for workmen required? yes __Yes__   ~~no~~ __$20.00/Day__

3. Railroad spur available? __No__   how near? ____

4. Working room none ____little ____ample __Yes__

5. Equipment rental available? __Yes__   how near? __10 miles__

6. Labor available? __Limited__   quantity? ____skill? __Average__

7. Subcontractors available? __Few. See remarks below__

8. Source of water __2" dia. pipe, 300 yards distant__

9. Source of power __Need transformer, overhead on site__ telephone __Available__

10. Need fences, barricades, lights, flagmen? __Fence only__

11. Soil conditions __Firm sandy clay__ hardness____wetness____

12. Extent of clearing, grubbing, trees, etc. __Grass & weeds__

13. Location of disposal area __5 miles__   fees __$50.00/load__

14. Source of import fill material __Not known - assume 4 miles__

15. Security requirements __Need watchman weekends__

16. Source of concrete, lumber, etc. __Within 15 miles__

17. Remarks __Not shown on drawings: concrete lined ditch 200' long x 6' wide; 20 pine trees 16" diam. average one conc. block building 20' x 40'. Local masonry subs. not available - advisable to budget the masonry work.__

Investigation made by __PC & CS__   Date __June 12, 1982__

Figure 4.4a

With the preceding information and additional notations made in red pencil on the drawings, we are ready to start the estimating work in earnest. The next step will be to list in proper detail, all categories of work in the project as they are customarily quoted by subcontractors.

Step 5 Summarize the specs

Figure 1.10a

By this time we have become familiar with the scope of the project in a shallow, overall way. Now, to go into greater depth let's lay the drawings aside and concentrate on the specs.

Using the *specification summary sheet* and beginning with the very first page of the specs, we list every trade for which sub-bids may be expected.

Specifications Summary Sheet

Project: Aircraft Parts Warehouse   Bid Date: July 7, 1982
Location Boulder, CA   Cal Days to Complete 600 Days

Sec	Trade	Base Bid	Alt A	Alt B	Alt C
01100	General conditions	X*			
01311	Network analysis	X*b*			
02110	Demolition	X*b*			
02200	Earthwork	X*b*			add*b*
02300	Prestressed conc piling	X			
02444	Chain link fencing	X			
02640	Asphalt concrete paving	X			·deduct
02800	Landscaping & sprinklers	X			
03300	C.I.P. concrete	X*		deduct*	add*
03400	Precast concrete	X*			
03600	Reinforcing steel	X			
04230	Concrete block masonry	X*b*			
05120	Structural steel	X			
05320	Steel roof decking	X			
05500	Miscellaneous metal	X			
	Install misc. metal	X*			
0600	Carpentry	X*			
07241	Roof insulation	X			
07510	Roofing	X			
07600	Sheet metal work	X			
07951	Caulking & sealants	X*			
08110	Hol. metal doors & frames	X			
08310	Sliding fire doors	X			
08330	Steel studs & gyp bd	X			
08710	Finish hardware	X			
09910	Painting	X			
11675	Pallet storage racks		add		
11871	Adjustable loading ramps			add	
15000	Mechanical	X			
16000	Electrical	X		add	
	Sub Total				

Figure 4.5a

If the specs are well and thoroughly written, this summary may prove to be complete; otherwise, a few additional items may later be discovered in the drawings.

This list is a skeleton outline of the estimate, and at the bid deadline a dollar amount will be written opposite each trade item.

Meanwhile, this list serves as a guide to the items which:

1. We will carefully estimate (*)
2. We will budget (b)
3. We will leave to sub-bidders

We will estimate general conditions

     C.I.P. concrete
     Precast concrete
     Installation of misc. metal
     Carpentry
     Caulking & sealants
     A portion of alternate bid B

We will budget

     Network analysis
     Demolition
     Earthwork
     Concrete block masonry

We will depend upon subcontract bids for all other work in the project.

Step 6 General conditions (outline)

We can only start the general (and special) conditions at this time. (See Figure 4.6a) The items and costs will develop as the estimating work proceeds and will be completed in the final stages, just before the bid deadline.

In this project, we list those items specifically named in the specs and those from our check list (Section 2.18) which we recognize as relevant. We immediately fill in known quantities and/or unit prices.

Section 2.18

Section 1.16

Before any pricing out can be done, we must establish the hourly pay rates for labor. The specified time to complete the construction is 600 calendar days. Although we may later compute and use a slightly longer or shorter period, the given 20 months will serve to establish our working pay rates.

The overwhelming majority of a general contractor's field employees are carpenters, cement masons and laborers. It is usually sufficient to confine our figuring to these three trades.

If hourly pay rates at the start of the project are:		At the project's completion they will be:	The average for estimating purposes will be:
Carpenter	18.13	21.94	19.65
Cem masons	17.29	20.92	18.74
Laborers	15.10	18.27	16.37

These increased levels are figured as in Section 1.16. Reference to carpenter union contracts shows the following projected increases during the 20 month construction period:

First 4 months	no change
Next 6 months	7% increase
Next 6 months	14% increase
Next 4 months	21% increase
	42%

42/2 = 21% net increase in 16 of the 20 months.

Adjust: 16:20::x:21 = 16.8% net increase for the entire project, and

16.8/2 = 8.4% the average increase. Although cement mason's and laborer's contracts differ somewhat from carpenter's, they are close enough that the same percentage increase may be used.

This average increase (8.4%) may be applied also to unit prices. Work to be performed at the project's start may be figured at the lowest hourly rate, or unit price; work to be performed near the finish may be figured at the highest rate, or unit price; otherwise, we will use the average of 8.4% increase.

We now lay aside the general conditions price-out sheet, along with the specifications summary sheet (step 5) for later completion.

Step 7 Demolition

Section 2.20

Section 3.1a

Figure 1.10f

Since our warehouse project is all new(no alteration demo work)this price-out is in the nature of a budget for comparison to sub-bids.

Demolition items, as shown on the drawings and described in the specs, are taken off and listed on the price-out sheet. A few additional items were observed at the site (step 4); these, too, are listed on the price-out sheet. All pricing out is done under the subcontract column, at the presumed bare cost; then an appropriate markup is applied. Theoretically, a subcontractor's bare cost would be lower than that of the general contractor, so that a markup would tend toward equality.

Circled numbers in the left margin refer to worksheets for detailed explanations.

Step 8 Earthwork

Section 3.1b

The terms earthwork, earthmoving, grading and rough excavation mean approximately the same thing. Choice is a matter of personal preference. Since both demo (site clearing) and earthwork are usually done by the same sub, we will show both categories on the same, or succeeding price-out sheets. Again, this is only a budget, as its purpose is to evaluate sub-bids, rather than replace or compete with them. If necessary, it can serve as a plug-in-figure until a valid sub-bid is received.

Earthwork items, as shown on the drawings, described in the specs and verified in the site inspection, are taken off, listed on the price-out sheet and priced out.

Typical cross section A-A, Figure 4.8a, shows that 3'-3" of fill will be required to raise the floor slab to the desired elevation above existing grade. Analysis gives the following information:

1. Following the demolition work, light stripping and disposal is required.
2. Some material will be excavated from the yard area and used as fill under the building.
3. Scarify and compact over the entire site.
4. Import fill material, then spread and compact it under the building area.
5. Fine grade over the entire site.

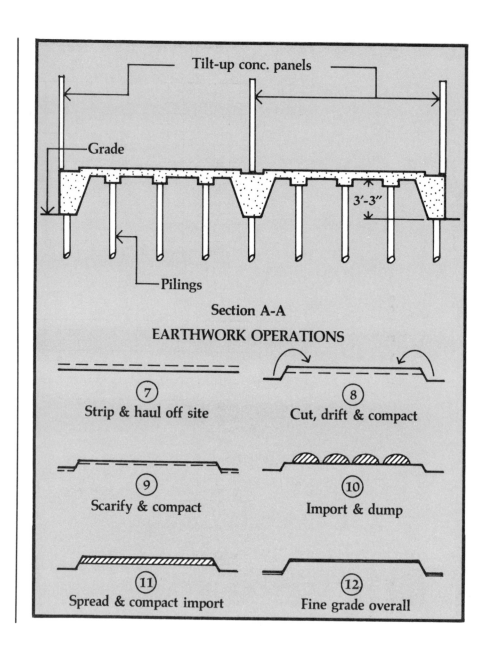

Figure 4.8a

PRICE OUT

JOB AIRCRAFT PARTS WAREHOUSE   **LOCATION** BOULDER, CA.   **BID DATE** JULY 7, 1982

FLOOR AREA _____ **COST PER SQ. FT.** _____   **DATE** JUNE 15, 1982   **SHEET** _____ **OF** _____

ITEM	DESCRIPTION		QUANTITY		MATL	LABOR	MATERIAL	SUB CONT.	TOTAL
	GENERAL CONDITIONS								
1	TRAVEL TO/FROM JOB (M&R)		MI		.40				
2	SUBSISTENCE - ALL EMPLOYEES		DAYS		20.00				
3	SURVEYING	FIG.2.18.B	CHR		132.00				
4	LAYOUT FOR STRUCTURES	FIG.2.18.D							
5a	FIELD OFFICE TRAILER	FIG.2.18.C	MO		77.22				
b	STORAGE SHED	FIG.2.18.E	MO		75.00				
c	TELEPHONE	FIG.2.18.F	MO		150.00				
d	CHEM. TOILETS (2 EA.)	FIG.2.18.G	LS	—	100.00				
e	WATER HOOKUP	FIG.2.18.I	LS	—					
f	ELEC. HOOKUP	FIG.2.18.I							
6	UTILITY MONTHLY CHARGES	FIG.2.18.I	MO		330.00				
7	OFFICE EQUIP. & SUPPLIES	FIG.2.18.J	MO		35.00				
8	SMALL TOOLS & RENTAL EQUIP.	FIG.2.18.M	M	5.5	.005				
9	SMALL VEHICLES, FUEL, ETC.	FIG.2.18.L	MI		.50				
10a	CLEANUP - PROGRESSIVE	FIG.2.18.N	HR						
b	FINAL		SF						
11	SUPERINTENDENT 21.65/HR		MO			3752.00			
12	CLERK 6.00/HR		MO			1040.00			
13	FENCE W/ 10' WIDE GATE	FIG.2.18.H	LF	600					
14	PROJECT SIGN		LS	—					
15	TEMP. CLOSURES FOR WEATHER								
16	WINDOW & FIXTURE CLEANING								
17	BUILDING PERMITS								
18	SPECIAL INSURANCE								
19	SCAFFOLDING		HR						
20	WATCHMAN WEEKENDS								
21	SAFETY RAILING ON ROOF								
	SUB TOTAL								
	FB 46%								
	TOTAL								

Figure 4.6a

189

PRICE-OUT

JOB __AIRCRAFT PARTS WAREHOUSE__ __July 7, 1982__ BID DATE _____ LOCATION __BOULDER, CA,__ DATE __June 15, 1982__ SHEET __2__ OF __7__

FLOOR AREA _____ COST PER SQ. FT. _____

ITEM	DESCRIPTION	REF	QUANTITY	LABOR	SUB	LABOR	MATERIAL	SUB CONT.	TOTAL
	DEMOLITION (SITE CLRG)								
1	Concrete lined ditch 4" thk.	Fig. 2.20.g	1200 SF		.35			420	420
2	Pine tree 16"∅ & stumps		20 ea.		159.09			3182	3182
3	One story conc. block bldg.	see worksheet	800 SF		1.72			1376	1376
4	slab & footings	Fig. 2.20.i	19 CY		98.32			1868	1868
5	Chain link fence 6' high	2.20.b	310 LF		1.30			403	403
6	Conc. gutter type curb	2.20.h	280 LF		2.07			579	579
7	saw cut concrete 1½" deep	2.20.a	20 LF		2.50			50	50
8	Conc. sidewalk 4" thk.	2.20.g	900 SF		.42			378	378
9	sewer m.h. 8' deep, 64cf	2.20.0	1 ea.		194²			194	194
10	A.C. paving 3" thk.	2.20.j	1500 SF.		.16			240	240
								8690	8690
	subcontractor mark-up	say	20%					1738	1738
									10428
	EARTHWORK								
1	Strip & haul offsite 4 miles		1900 CY		2.67			5073	5073
2	Cut, drift & compact on site		2844 CY		1.60			4550	4550
3	Scarify & compact overall		296,000 SF		.015			4440	4440
4	Import fill material		21,427 CY		4.37			93614	93614
5	Spread & compact the import		21,427 CY		.95			20350	20350
6	Fine grade overall		296,000 SF		.018			5328	5328
								133355	133355
	subcontractor mark-up	say	15%					20003	20003
									153358

Figure 4.7a

WORK SHEET

Figure 4.7b

JOB AIRCRAFT PARTS WAREHOUSE   **LOCATION** BOULDER, CA.   **BID DATE** JULY 7, 1982

FLOOR AREA ____   COST PER SQ. FT. ____   **DATE** JUNE 15, 1982   SHEET ____ OF ____

ITEM	DESCRIPTION		QUANTITY	LABOR	EQUIP. SUB.	LABOR	EQUIP.	SUB CONT.	TOTAL
	DEMOLITION								
1	PINE TREES - CUT & REMOVE	Fig.2.20.K				52		22	74
	STUMPS - EXC. & REMOVE					17		36	53
						69		58	127
	FB 46% OF LABOR								32
									159
2	CONC. BLOCK BLD'G DOWN TO SLAB								
	ROOFING MATERIAL	Fig.2.20.N	800 SF	—	.13			104	104
	ROOF SHEATHING & FRAMING	Fig.2.20.N	800 SF	—	.23			184	184
	CONC. BLOCK 8" UNGROUTED	Fig.2.20.M	960 SF	.55	.33	528	317		845
						528	317	288	1133
	FB 46% OF LABOR								243
									1376
3	CONC. FLOOR SLAB & FOOTINGS	Fig.2.20.i	19 CY	42.00	37.00	798	703		1501
	FB 46% OF LABOR								367
									1868
4	CONC. GUTTER TYPE CURB	Fig.2.20.H	280 LF	.97	.65	272	182		454
	FB 46% OF LABOR								125
									579
5	SAW CUT CONCRETE 1½" DEEP	Fig.2.20.A	20 LF	1.25				25	25
	DOUBLE FOR SMALL QUANTITY MINIMUM CHARGE							25	25
								50	50
6	SEWER MANHOLE 8" DEEP	Fig.2.20.0	57 CF	3.40				194	194

JOB __AIRCRAFT PARTS WAREHOUSE__ BID DATE __July 7, 1982__

FLOOR AREA _____ COST PER SQ. FT. _____ LOCATION __BOULDER, CA.__ DATE __JUNE 16, 1982__ SHEET ___ OF ___

ITEM	DESCRIPTION	QUANTITY	LABOR	SUB	LABOR	MATERIAL	SUB CONT.	TOTAL
	__EARTHWORK__							
⑦	Strip & haul 4 miles (1900 cy)							
	Excavator/loader (75/hr.)							
	3 trucks @ 55 (165/hr.)							
	240/hr.							
	Production: 10 cy/truck @ 3 cycles/hr =							
	30 cy X 3 Trucks = 90 cy/hr							
	240/90 = 2.67/cy	1900 cy		2.67			5073	5073
⑧	Cut, draft & compact on site							
	firm soil difficulty factor = .83							
	dozer/loader 80/hr.							
	compactor 50/hr.							
	watertruck 60/hr.							
	190/hr.							
	Production: allow 24 hr.							
	24x190 / 2844 = 1.60/cy	1.60/cy 2844 cy		1.60			4550	4550
⑨	Scarify & compact overall							
	dozer / compactor 90/hr.							
	watertruck 60/hr.							
	150/hr.							
	Production: 10,000 SF/HR.							
	150/10,000 = .015/SF	.015/SF 296,000 SF		.015			4440	4440
⑩	Import fill material							
	use quote from supplier	21,422 cy		4.37			93614	93614

Figure 4.8c

JOB __AIRCRAFT PARTS WAREHOUSE__   COST PER SQ. FT. _____

FLOOR AREA _____   LOCATION __BOULDER, CA__

BID DATE __JULY 7, 1982__   DATE __JUNE 16, 1982__   SHEET _____ OF _____

ITEM	DESCRIPTION	QUANTITY	LABOR	MATL	LABOR	MATERIAL	SUB CONT.	TOTAL
(11)	**EARTHWORK**							
	Spread & compact the import							
	Dozer 80/hr							
	Compactor 50/hr							
	Water truck 60/hr							
	190/hr							
	Production: 200 CY/hr							
	$\frac{190}{200}$ = .95/CY 21,422 CY			.95			20350	20350
(12)	Fine grade overall							
	Grader 65/hr							
	Production 3600 sf/hr							
	$\frac{65}{3600}$ = .018 296,000 sf			.018			5328	5328

Figure 4.8d

Step 9
C.I.P.
concrete
Section 2.21

We note two main footing types:

(1) *mass footings* integral with the floor slab. Form and brace one side. Considering the uniform height and the large quantity, consider fabricating 1/4 of the forms and using them 4 times. Note that interior footings and pile caps are merely extra thicknesses of the slab, and will require no forming.

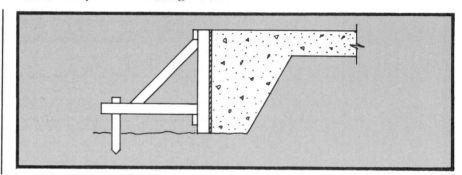

Figure 4.9a

Figure 2.21h

(2) *spread footing* at the office wing is similar to type 3; it is cast against the earth trench sides, and a template is provided for vertical rebar.

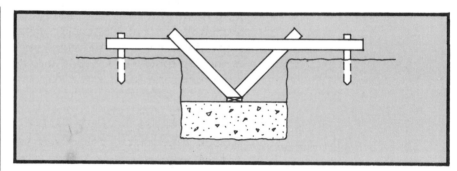

Figure 4.9b

In the price-out that follows, you will see a number of small items, marked (*) which have cost values, but are not standard enough for meaningful cost records so we either judge their costs, or obtain quotations from suppliers.

Because of the time required for site clearing, earthwork and pile driving, the formwork labor will be subject to the 8.4%, figured in step 6; however, the form materials will be purchased by pre-bid quotations (see worksheet item #15). For convenience, let us use the unit price from Table 2.21j, as the single formed surface of our mass footing is approximately the same as the face of a typical wall. Later we may wish to calculate its cost in greater detail.

In the pricing out, notice that all items are indented and grouped closely under their categories, as recommended in Sections 1.10 and 2.21. Also notice that the concrete and placing cost for mass footings is included with the slab work. This is a good example of the estimator's practice in choosing methods of construction.

Section 1.14

Foundation stem wall forming, 3' and higher, may be figured by values in Figure 2.21j (labor increased 8.4%).

Figure 2.21z

Figure 2.21cc

After forms are removed, wall faces will need pointing and patching, but only the portion which will ultimately be exposed to view will require architectural rub and grind, or sack finish.

Fine grading for the office wing slab may be figured from Figure 2.21c, but the warehouse slab is not typical and calls for a different approach (see #17 on worksheet).

Figure 2.21k

Columns poured between the precast wall panels will cost slightly more than freestanding columns because of the complication in securing and closing.

In this project, we will increase the tabulated labor values by 20%, plus the 8.4% figured in step 6 for pay increases; and we will increase the material and equipment by 10%.

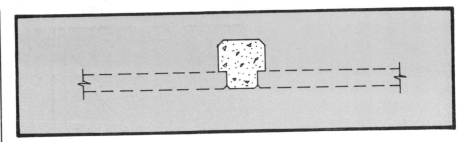

Figure 4.9c

Curb and walk costs are placed in the subcontract column for easy comparison with sub-bids, since these items are often quoted by specialists. If sub-bids are used later on, these budget figures may be crossed out.

Figure 3.1 d,e

In this project, we have two types of slabs on ground and we will price them separately. One slab is 12" thick in the warehouse area (see Section A-A) with extra thicknesses at footings and pile caps. The excavation for the mass footings has already been figured, but excavation for pile caps will be included with this slab work, as they require no forming. Also, the cost of removing the dirt left by the piling sub's drilling operations will be included with this slab work. As the concrete pouring will take several days, formwork will have to be constructed to end each day's pour, and to serve as construction joints. We will allow 20% of the concrete to be placed directly from trucks, and the remainder to be pumped into place.

Rebar needs no consideration, as it will be furnished and placed entirely by subcontractor.

The other slab is in the office wing. It is 4" thick and rests on two 2" layers of sand and 6 mil polyethelene membrane.

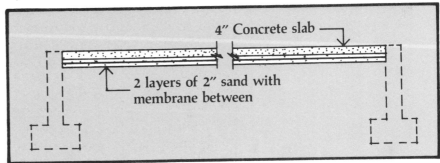

Figure 4.9d

We use the procedure as in Figure 2.21gg for pricing out.

Before going to the next step, let's do one last thing: check the cost of formwork materials. Adding items 2c, 2d, 3c, 4a, 5e and 9a we get 14,633 dollars in forming materials, as figured by the unit cost method; then, following the procedure in Figure 2.21r, we have the following comparison: **(see Figure 4.9e)**

Figure 2.21r

	Item	Gross Quant		Uses	Net Quant		Factor	Lumber Bf	Plywd Sf	Hdwe Factor	Hdwe $	Oil Gals
2c	Mass foot'gs	6,630	sf	4	2,427	sf	2.8	4,369	2,427	.02	133	33
2d	Depress for pls	2,040	lf	4	7471	lf	2.0	1,494	---	.02	41	10
3c	Spread foot'gs	380	lf	2	230	lf	2.25	518	---	.02	8	2
4a	Stem walls	2,280	sf	2	1,379	sf	2.67	2,303	1,379	.06	137	11
5e	Slab joints	6,480	lf	4	2,372	lf	2.00	4,744	---	.02	130	32
9a	Columns	6,120	sf	5	1,972	sf	2.7	3,352	1,792	.12	734	31
								16,780	5,778		1,183	119
	Unit Costs							.60	.50		---	5.0
								10,068	2,889		1,183	595
	Grand Total of Above								14,735			
	Total in Price-out								14,633			
	Difference								102			

Figure 4.9e

Carpenters 19.65 / hr.
Cement masons 18.74 / hr.
Laborers 16.37 / hr.

JOB __AIRCRAFT PARTS WAREHOUSE__ COST PER SQ. FT. _____

FLOOR AREA _____ LOCATION __BOULDER, CA.__ BID DATE __JULY 7, 1982__ DATE __JUNE 17, 1982__ SHEET _____ OF _____

ITEM	DESCRIPTION	REF.	QUANTITY		LABOR	MATL	LABOR	MATERIAL	SUB CONT.	TOTAL
13 1	Lay out, batter boards, etc.	fig. 2.21.a	Ls	—	—	—		1444		4846
2	Mass footings									
14	a. machine exc.	fig.2.21c	476	CY	—	3.33	—	1585		1585
	b. hand exc.		9,600	SF	.28	—	2660			2660
15	c. formwork		6,630	SF	2.28	.60	15116	3978		19094
16	d. form depression for p.c. pl.	*	2,040	LF	.57	.25	1163	510		1673
	E. disposal excess dirt		476	CY	—	2.53	—	1347		1347
	F. form const. jts. in footings	*	8	Ea	72.50	17.00	580	136		716
	g. Concrete (pour w/slab)									
	h. point & patch (200 sf/hr.)	fig. 2.21z	7300	sf	.095	.05	694	365		1059
	i. rub & grind (60 sf/hr.)	fig. 2.21cc	6000	sf	.31	.09	1860	540		2400
3	Spread footings									
	a. machine exc. (same as 2a.)		84	CY	—	3.33	—	280		280
	b. hand exc. (same as 2b)		950	SF	.28	—	266			266
	c. formwork- B 2 uses	fig.2.21.h	380	LF	2.33	.66	885	251		1136
	d. concrete 3000 psi-1" fig.2.21s & 2.21.t		31	CY	8.36	66.16	259	2051		2310
	E. disposal (same as 2E.)		45	CY	—	2.83	—	127		127
	f. backfill Condition D	Fig.2.21d	39	CY	8.18	3.76	319	146		465
4	Foundation stem wall									
	a. formwork P'high - 2 uses	fig.2.21J	2280	SF	2.70	.73	6156	1664		7820
	b. Concrete 3000 psi-1" fig.2.21.t 2.21s		31	CY	8.36	66.16	259	2051		2310
	c. point + patch 180 sf/hr.	fig. 2.21.z	2280	SP	.10	.05	228	114		342
5	12" thk. Slab on grd. in warehouse									
	a. machine Exc. pile caps see 2a		930	CY	—	3.33	—	3097		3097
	b. disposal from pile Caps see 2E		1990	CY	—	2.83	—	5632		5632
	c. hand exc. around piles 4 piles/hr.		1767	Ea	4.09	—	7227			7227
17	d. fine grade		156,000	SF	.04	.023	6240	3588		9328
	E. formwork @ joints 12": 4 uses	fig.2.21.i	4,480	LF	2.18	.42	14126	2722		16848
	f. set screeds	Fig.2.21.aa	152,000	SP	.112	.044	17472	6864		24336
18	g. Concrete 3000 psi fig.2.21.s + 2.21.t		8,400	CY	6.78	70.23	57036	589932		646968
	CARRY FORWARD						1359948	6284424		7643372

Figure 4.9f

PRICE OUT

JOB **AIRCRAFT PARTS WARE HOUSE** LOCATION **BOULDER, CA.** BID DATE **JULY 7, 1982** DATE **JUNE 18, 1982** SHEET ___ OF ___

FLOOR AREA ___ COST PER SQ. FT. ___

ITEM	DESCRIPTION	REF.	QUANTITY	LABOR	MATL	LABOR	MATERIAL	SUB CONT.	TOTAL
	C.I.P. CONCRETE		FORWARDED			1359948	6284924		7643372
	h. expan. jts & sealer 3/4"x11/2"	fig.2.21.ff	1560 LF	.78	.87	1217	1357		2574
	j. finish & cure	fig.2.21.bb	156,000 SF	.25	.025	39000	3900		42900
6	4" thick slab on grd. in office								
	a. fine grade 200 sf/hr	fig.2.21.c	12,000 SF	.08	—	960	—		960
	b. 1st 2" layer, sand mat'l	fig.2.21.ee	90 CY	11.65	11.65		1049		1049
	labor & equip.	fig.2.21.dd	90 CY	11.24	11.80	1012	1062		2074
	c. 6 mil poly membr.	fig.2.21.ee	13,200 SF	.033	.033	436	436		872
	d. 2nd 2" layer, sand mad'l	fig.2.21.ee	90 CY		11.65		1049		1049
	labor & equip.	fig.2.21.dd	90 CY	11.34	11.80	1012	1062		2074
	e. set screed	fig.2.21.aa	12,000 SF	.085	.033	1020	396		1416
	f. concrete 2500 PSI-1" fig.2.21.s +	2.21.t	166 CY	7.63	58.00	1267	9628		10895
	pumping	fig.2.21.t	133 CY	—	10.94		1455		1455
	g. finish & cure 600 sf/day	fig.2.21.bb	12,000 SF	.31	.025	3720	300		4020
7	Steps on grd.	*	120 LF	8.80	3.40	1056	408		1464
8	Sand fill & temp. slab in loading dock ramps *		3 Ea	262.33	145.00	787	435		1222
9	Columns between precast panels		—	—	—				
	a. formwork 18' high, 5 uses	fig.2.21.k	6,120 SF	4.40	.90	26928	5508		32436
	b. chamfers	*	8,160 LF	.25	.09	2040	734		2774
	c. concrete 3500 PSI-3/4" fig.2.21.s+fig.2.21.t		83 CY	8.86	74.90	735	6217		6952
	pumping 20CY/4 hr rate	fig.2.21.t	83 CY	40.66		3374	3374		3374
	d. point & batch 100 sf/hr	fig.2.21.z	6,120 SF	.19	.09	1163	551		1714
	e. rub & grind/sack 40sf/hr	fig.2.22.cc	6,120 SF	.47	.14	2876	857		3733
10	Pits for adj. loading ramps w/temp. fill	*	3 Ea	580.00	300.00	1740	900		2640
11	Curb, gutter type-straight	fig.3.1.Ea	350 LF	9.24				3234	3234
12	Curb, plain type-straight	fig.3.1.Ea	520 LF	8.49				4924	4924
13	Sidewalks	fig.3.1.Da	1550 sf	1.90				2945	2945
	Add rock salt	fig.3.1.Da	950 sf	.25				238	238
14	Catch basins 4'x4'x5'	fig.2.21.Y	2 Ea	617.00	308.00	1234	616		1850
	FB 467.	Fig.2.27a				224151	669718	11341	905210
									1031109
									1008319

Figure 4.9g

198

JOB __AIRCRAFT PARTS WAREHOUSE__   LOCATION __BOULDER, CA.__   BID DATE __JULY 7 1982__   SHEET ___ OF ___

FLOOR AREA _____ COST PER SQ. FT. _____   DATE __JUNE 17, 1982__

ITEM	DESCRIPTION	REF.	QUANTITY		LABOR	MAT'L	LABOR	MATERIAL	SUB CONT.	TOTAL
⑬	__C.I.P. CONCRETE__									
	Layout batter boards, etc.	fig.2.21.a								
	Corners		6	Ea	79.00	20.00	468	120		588
	Intersections		2	Ea	62.00	16.00	104	30		134
	Lines Continuous		2600	lf	.13	.10	325	250		575
	Pits + manholes		8	Ea	26.00	20.00	208	160		368
	Piling 40/hr = 44 hr	*	1767	Ea	1.30	.50	2297	884		3181
							3402	1444		4846
⑭	Machine Excav.									
	Backhoe 60/hr									
	Production 18 CY/hr 60/18 = 3 33/CY		474	CY	3.33		1585			1585
⑮	Formwork - 4'3" high	fig.2.21.j								
		Step 6	6630	sf	2.38	.60	15116	3978		19094
	Labor 2.10/sf² + 8.42₀									
	Mat'l .60/sf (no increase)									
⑯	Disposal	Sec.2.21	476	CY	2.83			1347		1347
	Loader 60/hr									
	Trucks 2@55 110/hr									
	170/hr									
	Production: 3 (10CY) loads/truck/hr 170/60									
⑰	Fine grade									
	6 men 8 days =		384	hr	16.37		6286			6286
	Small grader/tractor 8 days		64	hr	55.00		—	3520		3520
							6286	3520		9806

Figure 4.9h

WORK SHEET

Figure 4.9i

200

JOB __AIRCRAFT PARTS WAREHOUSE__ LOCATION __BOULDER, CA.__ BID DATE __JULY 7, 1982__ SHEET ____ OF ____

FLOOR AREA ____ COST PER SQ. FT. ____ DATE __JUNE 17, 1982__

ITEM	DESCRIPTION	REF.	QUANTITY		LABOR	MATL	LABOR	MATERIAL	SUB CONT.	TOTAL
⑱	_C.I.P. CONCRETE_									
	Concrete 12" slab									
	material 3000 PSI - 1½"	fig.2.21.5	8400	CY	—	63.00		529200		529200
	Equip 2 pumps 18 days @ 3250	fig.2.21.t	6720	CY	—	8.71		58531		58531
	labor 20 min 18 days		2880	hr	16.37	—	47146			47146
	foremen 4 ea. 18 days		576	hr	17.12	—	9861			9861
	Small tools + equip.		18	day	—	160.00		2160		2160
							57007	589891		646898

As it happens, the difference is only 102 dollars—not enough to require an adjustment; however, this procedure is a worthwhile practice, because in some projects, the difference might be considerable.

Step 10
Precast concrete (tilt-up) panels

Figure 2.22

Although no two precast concrete projects are the same, all projects have a basic method of construction in common. In order to price out our present unique project, we take off and tabulate the following pertinent facts:

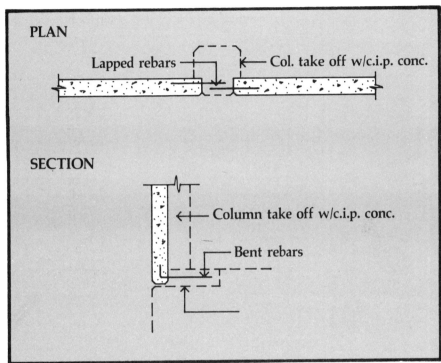

Figure 4.10a

Projecting rebar secures panels at edges and bottoms in c.i.p. concrete.

Because of bent bottom bars, avoid casting panels "pancake" fashion.

In this project, the floorslab is available for use as a casting slab, otherwise, a temporary casting slab would have been necessary.

Specific items and quantities are as shown on the following price-out sheet.

Prices for material and equipment items marked * are direct quotations from suppliers and are not to be found in the reference data given in this book.

Procedure for the construction of the wall panels is very much the same as for floor slabs, the main difference being a bond-breading compound on the casting slab and inserts for lifting, bracing, doorframes, anchoring of panels together, etc.

It is important to judge the number of panels to be formed and poured in a group, as a basis for figuring material re-uses, sizes of crews and scheduling of erecting equipment. In this project, we have 105 panels averaging 20' wide by 24' high and 7 1/2" thick. We judge a maximum daily concrete pour of 16 panels, or 80 cy of concrete; this works out at 6.58 re-uses of form material.

No consideration is given here to reinforcing steel (as it is a subcontract item), except the method of forming panel edges, where bars project, and casting where bent bars interfere.

PRICE OUT

JOB **AIRCRAFT PARTS WAREHOUSE**

FLOOR AREA _____ COST PER SQ. FT. _____ LOCATION **BOULDER, CA.**

BID DATE **JULY 7 1982**   DATE **JUNE 20 1982**   SHEET **5** OF **7**

ITEM	DESCRIPTION	QUANTITY	LABOR	MATL	LABOR	MATERIAL	SUB CONT.	TOTAL
	PRECAST (TILT-UP) CONCRETE WORK							
1	Engineering (drawing + calcs)	105 pls	—	*18.00	—	—	1890	1890
2	Bond breaker (796 SF/hr per man)	50,400 SF	.025	*.03	1260	1512		2772
(19) 3	Form edges (8 uses)	9,240 LF	1.84	.21	17002	2310		19312
4	Chamfer strips (110 LF/hr)	16,380 LF	.197	*.08	3227	1310		4537
5	Set bolts for roof ledgers (8/hr)	700 Ea	2.46	—	1722			1722
6	Set lift inserts (6/hr)	630 Ea	3.28	*3.95	2066	2489		4655
7	Set brace inserts (7/hr)	420 Ea	2.81	*3.33	1180	1399		2579
8	Set channel door frames (2 hrs each)	12 Ea	39.20	—	472			472
(20) 9	Concrete 3500 PSI 3/4" agg.	1290 CY	6.30	77.29	8064	100211		108275
10	Finish + cure (labor + 8.47%) Fig.2.21.bb	60,400 SF	.27	.025	13608	1260		14868
(21) 11	Erection- crane + riggers	105 pls	—	*347.00			36435	36435
12	Braces- install + remove (1½ hr each)	210 Ea	29.48	*20.00	6191	4200		10391
13	Leveling, grouting under panels (1½ hr each)	105 Ea	28.11	11.39	2952	1196		4148
14	Point + patch (1 hr Ea)	105 Ea	18.74	5.70	1968	599		2567
15	Scaffolding	105 hrs	1.00	1.00	105	158		263
					59817	116644	38325	214786
	FB 46% of labor							27616
								242302
	SUMMARY (compare to sub bids)							
	Casting- (items #1 thru #10)				48601	110491	1890	160982
	FB 46% of labor							22366
								183338
	Erecting- (item #11)				—	—	36435	36435
	Securing- (items #12 thru #15)				11216	6153		17369
	FB 46% of labor							5160
								22529
	Total Estimate Precast Concrete Work							242302

Figure 4.10b

JOB **AIRCRAFT PARTS WAREHOUSE**   LOCATION **BOULDER, CA.**   BID DATE **JULY 7, 1982**

FLOOR AREA _____ COST PER SQ. FT. _____   DATE **JUNE 20, 1982**   SHEET ___ OF ___

ITEM	DESCRIPTION		QUANTITY	LABOR	MAT'L	LABOR	MATERIAL	SUB CONT.	TOTAL
	PRECAST (TILT-UP) CONCRETE WORK								
19	Form panel edges	fig.2.21.i	9240 LF	1.24	.25	17002	2310	—	19312
	Cost of 8", 4 uses would be labor								
	1.70/LF; mat'l .29/LF								
	for 6.58 uses use formula	fig.2.4.e							
	$\frac{KWP}{U} = \frac{1.67 \times 2 \times .5}{6.58} = .25/LF$	fig.2.4.g							
	labor 1.70 ÷ 8.47% = 1.84/LF								
20	Concrete 3500 PSI 3/4" agg. - mat'l.	fig.2.21.s	1280 CY	—	74.90	—	95872	—	95872
	labor @ 16 pours = $\frac{1280}{16}$ = 80 CY/day	fig.2.21.t	1280 CY	6.30	—	8064	—	—	8064
	= 5.81/hr + 8.47% = 6.30/CY								
	Equip @ 10.16/CY - ⅓ placed by		1280 CY	—	3.39	—	4339	—	4339
	pumping = $\frac{10.16}{3}$ = 3.39/CY						8064	100211	—	108275
21	Erection - crane+riggers @ ¾ hr/panel	fig.2.10.a							
	Ave. wt. 22 tons/pl.		150 hr	—	120.00	—	18000	—	18000
	Use 45 ton crane .75 + 10 hrs move in/out	$\frac{}{.105}$							
	Riggers 3@ 140hrs		420 hr	—	32.42	—	13616	—	13616
	Rigger foreman		140 hr	—	34.42	—	4819	—	4819
							36435	—	36435

Figure 4.10c

After all inserts are placed, concrete poured, finished and cured, the panels are ready for tilting up and bracing. We are fortunate in this project, that the panels need not be cast in a location such that conveying and awkward maneuvering by crane is necessary. However, there is a problem due to the projecting bent bars at the panel bottoms. If the panels are cast with inside faces down (for ease of tilting up) the bars may have to be bent 180 degrees, set in the slab depression, and the depression filled with sand for casting the concrete, Figure 4.10d.

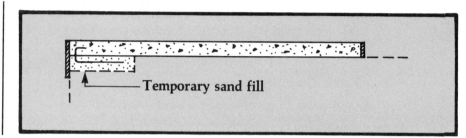

Figure 4.10d

After the panels are tilted up, the sand is removed and the bars are bent back to 90 degrees. If, on the other hand, the panels were cast with their inside faces upward, they would have to be rotated by crane, requiring more crane time and risking breakage.

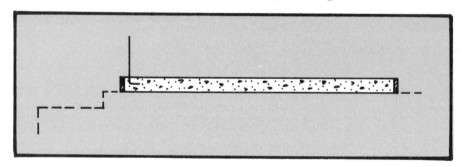

Figure 4.10e

For estimating purposes, we choose the second method, and the extra crane time is included in the 3/4 hour per panel as figured. This is a good example of the estimator's duty to select methods of construction.

Section 1.14

Section 1.19

Within this problem lies the possibility for a future value engineering proposal, to be discussed at the upcoming presentation (hashing-over) meeting. A change of design as follows may be offered to the A & E for their consideration, Figure 4.10f.

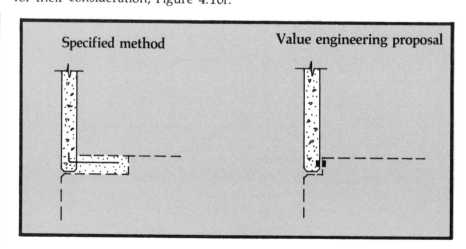

Figure 4.10f

204

Provide welding inserts in the floor slab and panels, thus eliminating the bent bar casting problem, the grouting over the bent rebar, and the patch-work concrete finish.

The payment to riggers is included in the equipment, rather than the labor column, because riggers are not usually hired on the payroll, but are furnished, along with the crane, at gross "rental" rates.

The work of leveling and grouting under the panels proceeds with the pace set by the tilting up operation.

The pointing and touching up is an independent operation, and since it involves scaffolding, would be more economically accomplished after the c.i.p. columns are poured, and the braces removed.

Figure 2.22a

The cost items may now, if desired, be re-grouped for comparison to possible future sub-bids.

This completes the estimating of all the concrete work in our project, and is a good point from which to look backward over what we have done and forward to what we have yet to do.

Step 11 Backward/ forward look

A Backward Look

Step 1
We have tentatively rated the project at 88 on the desirability scale.

We are four days ahead of our estimating time schedule, but complacency is not advised; unforeseeable interruptions could cancel out this lead time or another desirable project might come along for consideration in our overall schedule.

Step 2
We have tentatively analyzed the bid schedule.

Step 3
Step 4
We have inspected the job site and recorded important data affecting costs and procedures.

Step 5
We have summarized the specs and made up a specification summary sheet, which serves to guide our decisions regarding which items to (1) estimate, (2) budget, or (3) rely upon sub-bids.

Step 6
We have only outlined the general conditions (for future completion), and established the level of the main trades' pay rates to be used in this project; and we have figured the average percentage (8.4%) which must be applied to all pay rates and labor unit prices current at the start of this project.

We have taken off and budgeted:

Step 7
Demolition

Step 8
Earthwork

We have taken off and estimated:

Step 9
C.I.P. concrete

Step 10
Precast concrete

A Forward Look

We have yet to take off and estimate:

Installation of miscellaneous metal
Carpentry
Caulking & sealants

We have to complete the general conditions, with the help of a progress chart and accumulated knowledge of the project working conditions.

We have to budget:

Network analysis
Masonry

We have yet to estimate portions of the alternate bids.

We have yet to present our work to our company's management staff, and make any alterations that may result.

We have yet to incorporate any addendum changes, write in the sub-bids, apply a markup and bid the job.

Step 12 Install miscellaneous metal

Section 2.25

The items and quantities are listed on the price-out sheet as they are taken off. The unit prices are derived from the estimator's judgment based upon experience; accurate cost records are impossible in work so varied as the setting of miscellaneous metal. Yet, nearly every project has miscellaneous metal to install, so the estimator acquires a "feel" for their cost values.

None of these items are so extremely heavy as to call for individual equipment pricing; therefore, a single item of allowance is made to cover all equipment (cranes, etc.) for this category of work.

Another reason that the cost of installing miscellaneous metal is only approximate is that sub-bidders vary in the items which they propose to install, or furnish F.O.B. jobsite.

In this pricing out, we figure those items which we think will *probably* be furnished F.O.B. jobsite. Often a last minute adjustment is in order at the bid deadline.

Step 13 Carpentry

Section 2.23
Section 2.24

In this project, both rough and finish carpentry are minor trades, so they are combined for pricing out.

The cabinet work may be too trivial to arouse millwork subcontractors' interest, so we budget its cost at this time.

Because of the small quantities, production will be low and unit costs high.

Step 14 Caulking and sealants

Section 2.26

In some projects, caulking becomes complex and very costly. In this project, it is a small cost item; but it is always advisable to trace all details that may require caulking, as it is very easy at a casual glance to underestimate the quantity and the value. In recent years, caulking has increased in importance as a distinct construction trade.

Step 15 Masonry

Section 3.1i

In the jobsite investigation, we noted that masonry subcontractors were not available locally; so in order to off-set a possible no-bid situation, or a single bid of unknown validity, we should place our own value on the work.

The takeoff shows a straightforward job of 8″ solid grouted, fluted, integral color block, with seven openings requiring temporary shoring.

Because a masonry contractor must pay the cost of travel and subsistence to this project, we add an allowance of 10% to the budget subtotal.

Figure 4.12a

JOB AIRCRAFT PARTS WAREHOUSE BID DATE JULY 7, 1982

FLOOR AREA _____ COST PER SQ. FT. _____ LOCATION BOULDER, CA. DATE JUNE 21, 1982

ITEM	DESCRIPTION		QUANTITY	MATL	LABOR	LABOR	MATERIAL	SUB CONT.	TOTAL
	INSTALL MISC. METAL								
1	Pipe guard posts incl. exc. + conc.		8 Ea	15.00	16.37	131	120		251
2	Angle door sills	½ mhr	12 Ea	—	9.83	118			118
3	Channel door frames (included in precast conc. work)			—	—				—
4	Angle in edge of dock	25 Lf/mhr	340 Lf	—	.79	269			269
5	Pipe sleeves for handrail	¼ mhr	6 Ea	—	4.94	29			29
6	Dock bumpers	1 mhr	30 Ea	—	19.65	590			590
7	Angle in adj. loading ramp pits	10 Lf/mhr	196 Lf	—	1.97	386			386
8	Ladders to roof	2 mhr	3 Ea	—	39.20	118			118
9	Roof scuttles	6 mhr	4 Ea	—	117.90	472			472
10	Catch basin frames + covers	¾ mhr	2 Ea	—	14.74	29			29
11	Safety nosings on steps	10 Lf/mhr	24 Lf	—	1.97	47			47
12	Anchor plates in tops of cols.	⅔ mhr	72 Ea	—	13.17	948			948
13	Ledger angles	25 Lf/mhr	380 Lf	—	.79	300			300
14	Misc. hoisting equipment		20 hr	65.00	—		1300		1300
						3437	1420		4857
	FB 46% on labor								1581
									6438
	CAULKING & SEALANTS	fig. 2.26 c							
1	Metal door frames - latex	½ x ½	80 Lf	.12	.52	42	10		52
2	Tilt-up concrete control jts.	¾ x ¾	240 Lf	.23	.59	142	55		197
a	Backer rods ¾"	fig. 2.26.b	240 Lf	.52	.39	94	125		219
3	Masonry wall control jts.	¾ x ¾	80 Lf	.23	.59	47	18		65
4	Gypsum board - Elastomeric	¼ x ¼	300 Lf	.03	.46	138	9		147
a	Backer rods ¼"	fig. 2.26.b	96 Lf	.18	.36	35	17		52
5	Miscellaneous - Polysulphide	¼ x ½	150 Lf	.23	1.12	168	35		203
						666	269		935
	FB 46% on labor								306
									1241

ITEM	DESCRIPTION		QUANTITY	LABOR	MAT'L	LABOR	MATERIAL	SUB CONT.	TOTAL
	CARPENTRY								
1	Metal door frames, grouted	2 hr ea	9 Ea	39 ^{30}	4 ^{00}	354		36	390
2	Metal doors 3'X7' S.C.	fig.2.24.c	4 Ea	27 ^{22}	—	109			109
3	Wood doors 2⅝X7⁰	fig.2.24.b	5 Ea	34 ^{82}	—	174			174
4a	Door hardware - thresholds	fig.2.24.d	2 Ea	19 ^{45}		39			39
b	closers	"	4 Ea	59 ^{05}		236			236
c	exit devices	"	2 Ea	78 ^{61}		157			157
d	kick plates	"	2 Ea	14 ^{74}		29			29
e	weather strip	"	2 Ea	39 ^{31}		79			79
5a	Nailers on roof 3X4, 2X6	18 BF/hr	180 LF	1 ^{09}	.90	196	162		358
b	3X8, 2X12	16 BF/hr	360 LF	1 ^{23}	1.10	443	396		839
6	Base cabinet - std. grd.	fig.2.24.e	8 LF	9 ^{84}	*70 ^{00}	79		560	639
7	Overhead wall cabinet	"	20 SF	2 ^{11}	*15 ^{00}	42		300	342
8	Misc. rough hardware		LS		LS		75		75
	FB 46% on labor					1937	669	860	3466
									891
									4357
	MASONRY								
1	Concrete block 8"sol. grout	fig.3.11a	3660 SF		6 ^{16}			26856	
a	Add for fluted				1 ^{10}			840	
b	Add for color				.20				
	Total				7 ^{46}			27696	
c	Add for shoring in windows/doors		7 Ea	*120 ^{00}				840	
	Subtotal							27696	
d	Add milage for out-of-town sub			*10%				2770	
								30466	30466
	NETWORK ANALYSIS								
1	fig.2.18 p 5,500,000 $				.001		55000		5500

JOB **AIRCRAFT PARTS WAREHOUSE** FLOOR AREA ___ COST PER SQ. FT. ___ LOCATION **BOULDER, CA.**

BID DATE **JULY 7, 1982** SHEET **7** OF **7** DATE **JUNE 21, 19 82**

Figure 4.13a

208

Step 16
Network analysis

Section 2.19

Figure 2.18p

Figure 2.19b

Our project specs require a network analysis to guide and control the time span from start to finish of the construction work. We will include in our estimate the cost of preparing and periodically updating the network according to the formula given in Section 2.18. For a simple project of 5,500,000 dollar size, the factor is .001.

However, at this time we need to make a simple progress schedule in bar graph form to help us decide the time span for use in our general conditions estimate, which was left unfinished in step 6. For its outline, let us use the summary sheet made in step 5. As we construct the graph, let us avoid extremes of optimism and pessimism, and try to be as realistic as possible.

Move-on includes such items as: temporary office, utilities, surveying and layout work. Clearing and earthwork time is easily approximated from the items and quantities given. It is evident that the earthwork must be substantially completed before the piling can be driven, and most of the piling must be completed before concrete formwork can begin. It is just as well that such delay exists, in order that shop drawings may be approved for piling, reinforcing steel and formwork. After 60 days pile driving can start, and if 44 piles can be driven each day, then 1767/44 = 40 working days, or eight weeks, will be required. Our price-out sheets show footings and slabs to progress at an average of about 435cy per day, and 8711/435 = 20 working days, or 4 weeks. Allow four weeks additionally for preliminary excavation, form work, finishing and curing, for a total of eight weeks. Reinforcing steel work proceeds throughout this period and that of the succeeding precast concrete work. Notice that some minor concrete and rebar work—walks, curbs, paving, etc., takes place near the end of the project.

The precast wall panels follow the floor slab work. If we allow two days between each of the 16 pours, to form and to set rebar, the casting time will be 16 + 30 = 46 days. The estimated erecting time is 150/8 = 19 days; but approximately half—say 10 days of this may occur during the period of casting (erect cured panels in increments as they are made). Net time for precast work is 46 + (19-10)/5 = 11 weeks.

Masonry work can start as soon as cast-in-place concrete is finished, and should take about three weeks.

Structural steel work cannot make progress until the precast concrete work is completed.

Miscellaneous metal work stretches over a long period, beginning with items which embed in concrete.

Metal roof deck follows the structural steel. Roofing, sheet metal and miscellaneous trades then proceed parallel and independent of each other.

Mechanical and electrical work begin with underground and rough-in soon after the start of the project, and continue through to the end of the project. With final trades—paving, fencing and landscaping, the time for completion of the project reaches 18 months, which is two months less than the time required by the specs.

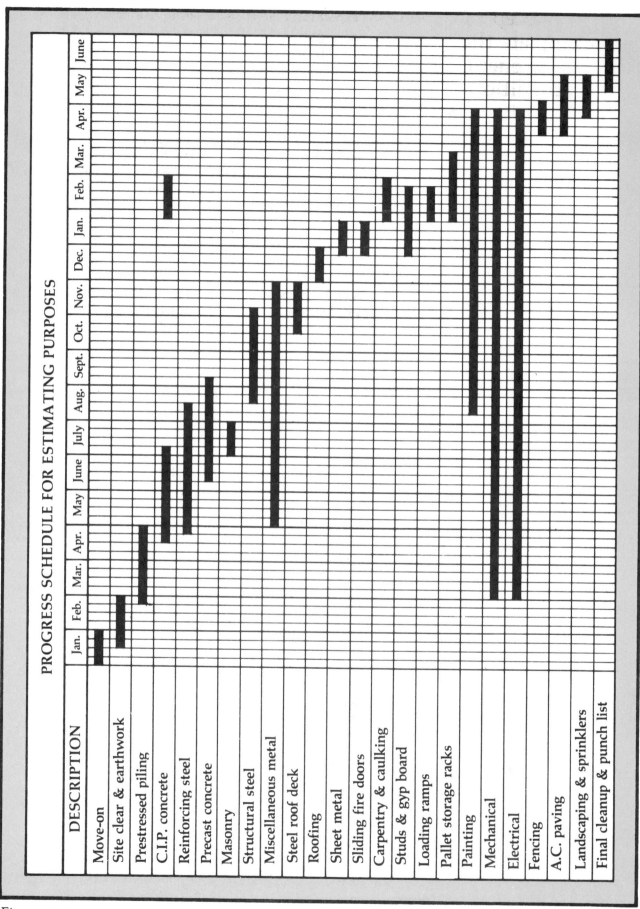

PROGRESS SCHEDULE FOR ESTIMATING PURPOSES

DESCRIPTION	Jan.	Feb.	Mar.	Apr.	May	June	July	Aug.	Sept.	Oct.	Nov.	Dec.	Jan.	Feb.	Mar.	Apr.	May	June
Move-on																		
Site clear & earthwork																		
Prestressed piling																		
C.I.P. concrete																		
Reinforcing steel																		
Precast concrete																		
Masonry																		
Structural steel																		
Miscellaneous metal																		
Steel roof deck																		
Roofing																		
Sheet metal																		
Sliding fire doors																		
Carpentry & caulking																		
Studs & gyp board																		
Loading ramps																		
Pallet storage racks																		
Painting																		
Mechanical																		
Electrical																		
Fencing																		
A.C. paving																		
Landscaping & sprinklers																		
Final cleanup & punch list																		

Figure 4.16a

210

We should expect the general conditions to cost approximately:

5,500,000 x .039		214,500
Special conditions	travel	5,299
	subsistence	57,811
	special ins.	6,875
	watchman w/fb	30,275
	Total	314,760

In step 4 we found by actual odometer measurement the round trip distance from main office to jobsite is 92 x 2 = 184 miles. If one inspection trip per week for 18 months is made, the mileage would be: 184 x 4 x 18 = 13,248. The unit price of .40/mile is our company's valuation of the gross cost for automobile transportation.

In step 4 it was found that subsistence at 20/day per workman is required at the site of our project. We may figure the total bare labor cost for our own company's work (subs are responsible for their own subsistence) from the price-out sheets:

General conditions	132,019
C.I.P. concrete	224,151
Precast concrete	59,817
Carpentry	1,937
Install misc. metal	3,437
Caulking & sealants	666
Total	422,027

From step 6 we find the average pay rate of our three main trades is:

$$\frac{19.65 + 18.74 + 16.37}{3} = 18.25/hr, \text{ and}$$

18.25 x 8 = 146/day. Total subsistence cost is $\dfrac{422,027 \times 20}{146} = 57,800$

For surveying, allow:

Property lines & corners	4 hrs
Rough grading stakes	16 hrs
Corners of structures	4 hrs
Finish grade stakes	6 hrs
Paving, walks, curbs	6 hrs
Utilities & misc.	4 hrs
	40 hrs

Figure 2.18b

To figure layout for structures:

Footing excavation (in C.I.P. concrete)	—	—		—	—
Sidewalks	1550 sf	@	.05	=	78
Curbs	830 sf	@	.09	=	75
Paving	12000 sf	@	.04	=	480
Catch basins, misc.	5 ea	@	34.0	=	170
					803

Figure 2.18d

Water hookup, meter and temporary piping is usually done by our own employees; electrical hookup, meter, pole and service wiring is usually done by negotiation with the electrical subcontractor. Here, we can only make an allowance based upon similar costs on other projects.

For small tools and miscellaneous rental equipment combined, we decide to use the formula suggested in Figure 2.18m.

For mileage of small vehicles (pickup trucks, etc.) in and about the project, we figure 20 working days per month and 20 miles per day: $18 \times 20 \times 20 = 7,200$ miles.

Figure 2.18n   We figure the cleanup work as follows:

Progressive	structure	@ 2 mhrs/day		
		$18 \times 20 \times 2 \times$	$16.37 =$	5,238
	site	120,000 sf $\times$	.03 =	3,600
				8,838
Final	structure	180,000 sf $\times$	.06 =	10,800
	site	120,000 sf $\times$	.02 =	2,400
				13,200

Figure 2.18b   The budget for a temporary fence around the construction yard is as follows:

Fence	600 lf @ 1.80	=	1,080
Gate	1 pr ls	+	96
			1,176

Temporary closures for weather can be an important item in some projects; here, it is minor, in the office wing only. Let us use an allowance for labor of six man hours per week in the final four months only, and a lump sum of 800 for material.

$$\text{Labor will be } 6 \times 4 \times 4 \times 19.65 = 1,886$$

The costs of window and fixture cleaning, building permit fees and special insurance are found by inquiry of business and municipality sources.

The total estimated cost of general conditions (321,338) is confirmed by the amount predicted at the start of this section.

Step 18 Estimate the alternate bid items

Step 3
Section 2.27

Alternate bid A is entirely subcontract work and requires no estimating by us.

Alternate bid B is item 8 on the cast-in-place concrete price-out sheet, plus its share of fringe benefits on labor.

Bare cost labor and material	1,222
FB 46% of labor (787)	362
	1,584 deductive

Alternate bid C requires the budgeting of concrete paving, which we do as follows. The takeoff gives us 12,000 square feet of 8" thick concrete on a 6" thick aggregate base course. Section 3.1C shows two methods for budgeting concrete paving (1) detailed method, and (2) unit price method. Because we expect paving sub-bids, great accuracy is not necessary here, so we will use the unit price method. Note that extra excavation is required due to the concrete being thicker than

Figure 3.10Ca   AC.

8" concrete paving	12,000 sf @	2.66 =	31,920
6" agg base course	266 cy @	20.00 =	5,320
Exc & haul off site	188 cy @	2.67 =	502
			37,742 additive

PRICE OUT

JOB **AIRCRAFT PARTS WAREHOUSE**   LOCATION **BOULDER, CA.**   BID DATE **JULY 7, 1982**   DATE **JUNE 23, 1982**

FLOOR AREA _____ COST PER SQ. FT. _____   SHEET _____ OF _____

ITEM	DESCRIPTION	REF.	QUANTITY	LABOR	MATL	LABOR	MATERIAL	SUB CONT.	TOTAL
	GENERAL CONDITIONS								
1	Travel To/From Job (M&B)		13,248 MI	—	.40		5299		5299
2	Subsistence - All Employees		2,890 Days	—	20.00		57800		57800
3	Surveying	Fig.2.18.B	40 CHR	—	132.00		5280		5280
4	Layout For Structures	Fig.2.18.D	LS	—	—	803			803
5a	Field Office Trailer	Fig.2.18.C	18 Mo	—	77.22		1390		1390
b	Storage Shed	Fig.2.18.E	18 Mo	—	75.00		1350		1350
c	Telephone	Fig.2.18.F	18 Mo	—	150.00		2700		2700
d	Chem. Toilets (2 Ea.)	Fig.2.18.G	18 Mo	—	100.00		1800		1800
e	Water Hookup	Fig.2.18.I	LS	—	—	300	200		500
f	Elec. Hookup	Fig.2.18.I	LS	—	—			1800	1800
6	Utility Monthly Charges	Fig.2.18.I	18 Mo	—	330.00		5940		5940
7	Office Equip. & Supplies	Fig.2.18.J	18 Mo	—	35.00		630		630
8	Small Tools & Rental Equip.	Fig.2.18.M	5.5 M	—	.005		27500		27500
9	Small Vehicles, Fuel, Etc.	Fig.2.18.L	7200 MI	—	.50		3600		3600
10a	Cleanup - Progressive	Fig.2.18.N	LS	—	—	8838			8838
b	Final		LS	—	—	13200			13200
11	Superintendent 21.65/HR		18 Mo	3752.00	—	67536			67536
12	Clerk 6.00/HR		18 Mo	1040.00	—	18720			18720
13	Fence w/10' Wide Gate	Fig.2.18.H	600 LF	—	1.96			1176	1176
14	Project Sign		LS	—	—			500	500
15	Temp. Closures For Weather		LS	—	—	1886	800		2686
16	Window & Fixture Cleaning		LS	—	—			1200	1200
17	Building Permits		5.5 M	—	500.00		2750		2750
18	Special Insurance		5.5 M	—	1250.00		6875		6875
19	Scaffolding (Included In Concrete Work)			—	—				—
20	Watchman Weekends		18 Mo	1152.00	—	207360			20736
	Sub Total					132019	123914	4676	260609
	FB 46% On Labor								60729
	Total								321338

Figure 4.17a

Step 19
Reappraise desirability rating/determine ideal markup

Figure 3.10a

Step 16

In step 1 we gave the project a rough appraisal of 88 for desirability. Now let's look at it again from the advantage of our greater in-depth knowledge. Using our own format we list the following judgment values:

1.	Location	90 miles	50
2.	Character of the project	Simple	90
3.	Quality of drawings & specs	Mediocre	80
4.	Quality of supervision	Excellent	90
5.	Quality of workmen	Good	85
6.	Cooperativeness of A. & E., etc.	Good	85
7.	Completion time	Ample	90
8.	Proportion subbed out	Good	80
9.	Need for work	Fair	80
10.	Economic conditions	Poor	100

(poor on masonry)

Sub Total 830

÷ 10

Rating 83

Figure 3.10b
Section 3.5

This is less optimistic than our earlier appraisal.

But let us now apply this rating to the markup finder scale. A line drawn from 5.5 million project size to 83 desirability rating intersects the markup scale at 9% (overhead and profit). This would be the ideal markup, and would amount to 5,500,000 x .09 = 495,000. It should be noted that the *ideal* markup may be subject to a judgmental adjustment at the final moment of bidding due to facts which may emerge. See more on this in the section on contingency allowances.

Step 20
Presentation meeting

Section 3.6
Step 1 & 19
Step 4

Step 5

Step 6
Steps 7 & 8
Step 9

Step 10

We are now ready to present our work to the supervisory and managerial staff for criticism and/or approval. Let us approach this meeting with the determination to be our most severe critic. Here is the outline and list of topics for explanation and discussion:

1. The desirability rating and its influence on the level (looseness/tightness) of the estimated costs.
2. General impressions and specific details that were noted during the jobsite investigations.
3. Evaluation of subcontractor trades for their relative importance and the probable competitiveness of the bidders. We will propose that particular attention be given to the following trades for solicitation of bids:

Prestressed concrete piling
Reinforcing steel
Masonry
Structural steel & roof decking
Roofing & roof insulation
Sliding fire doors
Pallet storage racks
Mechanical work
Electrical work

4. The level of pay scales at the beginning, at the ending of the project, and the average to be used in the unit prices.
5. Budgets made for demolition and earthwork.
6. Cast-in-place concrete, item by item and detail by detail including methods for excavating, forming, placing and finishing the concrete.
7. Precast (tilt-up) concrete work including methods of casting, erecting and securing in place.

8. At this point we will suggest for a future value engineering proposal the substitution of welding inserts for the specified projecting rebars at the bottom of panels.

While on the subject of value engineering, we will backtrack for a moment to cast-in-place concrete and propose for consideration a change in pile caps as follows:

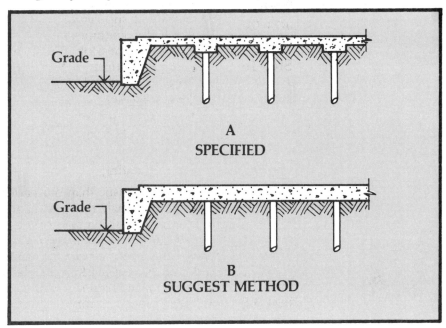

Figure 4.20a

Simply thicken the floor slab uniformly, as in sketch B, to avoid the excavating and forming of individual pile caps.

Step 12 10. Miscellaneous metal installation.
Step 13 & 14 11. Carpentry and caulking work.
Step 15 12. Masonry budget.
13. Network analysis budget, and the bar chart which indicates that the construction time may be estimated at 18 months rather than the
Step 16 specified 20.
14. Show how the total cost of the general conditions is confirmed by
Step 17 the formula in Figure 2.18a.

Present the reasoning in each item of the general conditions and make revisions as indicated by the consensus of opinion.

15. Since the alternate bids are simple, give little time to their presentation.
Figure 2.27a 16. Explain how the fringe benefits (and payroll taxes) are determined for this project.
17. Explain the proposed markup as determined by the formula in
Step 19 Figure 3.10b.
Figure 1.10b 18. Discuss any unresolved items on the notes and question sheet.

**Step 21
Bid
the
job**
Figure 1.6a
Step 5

Let us now gather together all of the price-out sheets, number them consecutively beginning with general conditions, and make the adjustments resulting from the presentation meeting. We will retain the original figures by simply crossing through and printing the revised figures over them.
Taking the specifications summary sheet, let's fill in the trades we have estimated and budgeted.

Next, let us enter those sub-bids which we determine by analysis and comparison to be the most complete and lowest in price (with risks taken into account). In the process we may use a few plug-in-figures, but before the final moment of bidding we should replace as many of the budget and plug-in-figures as possible with firm sub-bids or estimates.

Figure 3.9a

Figure 3.4a

The estimate summary sheet has now become the *bid sheet*. At its completion, we have in our base bid six estimates (general conditions, carpentry, cast-in-place concrete, precast concrete, installation of miscellaneous metal, and caulking) and two budgets (network analysis and masonry). All other figures are sub-bids.

The main apparent risk is the masonry trade, for which there were no sub-bids. Our company's own direct financial burden is as follows:

General conditions	309,120
C.I.P. concrete	972,931
Precast concrete	242,302
Install misc. metal	6,438
Carpentry	4,357
Caulking	1,241
Total	1,536,389

This is approximately 30% of the total cost of the project, which experience indicates is higher than average (we should have expected about 20%). Here is how we may rationalize this and adjust for it:

Step 19

Figure 3.10b

Section 3.5

In the desirability rating analysis, 80 had been given to the proportion of trades subbed out. Now we change our value to 60, which lowers the net total from 83 to 81 and raises the markup from 9% to 9 1/2%, for an additional 25,000 dollars. But in the presentation meeting, the consensus had been to cut the markup to 8 1/2%, so let us include this entire business in a contingency allowance, rationalized as follows:

Risk in subcontract bids (say 10% of masonry)	+ 3,000
Slack in certain items, approximated	-22,000
Hidden allowances	-8,000
Reduce markup by ½% (per presentation meeting)	-25,000
Net cut	52,000
Less contingency on our own work, as computed above	25,000
Reverse contingency allowance (cut)	27,000

Figure 3.10c

Finally, we must add the cost of bond. We find that the decimal for projects ranging from 5 to 6 million dollars is .005, or 1/2%. With this cost computed and added on, our estimate has now metamorphosed into a bid, or more precisely, the amount which we are willing to offer in the form of a bid.

PRICE OUT

JOB __AIRCRAFT PARTS WAREHOUSE__   BID DATE __JULY 7, 1982__   SHEET __1__ OF __7__

FLOOR AREA _____   COST PER SQ. FT. _____   LOCATION __BOULDER, CA.__   DATE __JUNE 23, 1982__

ITEM	DESCRIPTION	REF	QUANTITY	LABOR	MATL	LABOR	MATERIAL	SUB CONT.	TOTAL
	GENERAL CONDITIONS								
1	Travel To/From Job (M&R)		13,248 MI	—	.40		5299		5299
2	Subsistence - All Employees		2,890 DAYS	—	20.22		57800		57800
			32 / 40 CHR				52240		52240
3	Surveying	Fig.2.18.B	LS	—	—	803			803
4	Layout For Structures	Fig.2.18.D	LS	—	—	803			803
5a	Field Office Trailer	Fig.2.18.C	18 Mo	—	77.22		1390		1390
b	Storage Shed	Fig.2.18.E	18 Mo	—	76.00		1350		1350
c	Telephone	Fig.2.18.F	18 Mo	—	150.00		2700		2700
d	Chem. Toilets (2 EA.)	Fig.2.18.G	18 Mo	—	100.00		1800		1800
e	Water Hookup	Fig.2.18.I	LS	—	—	300	200		500
f	Elec. Hookup	Fig.2.18.I	LS	—	—			1800	1800
6	Utility Monthly Charges	Fig.2.18.I	18 Mo	—	275.00 / 330.00		4950 / 5940		5940
7	Office Equip. & Supplies	Fig.2.18.J	18 Mo	—	35.00		630		630
8	Small Tools & Rental Equip.	Fig.2.18.M	5.5 M	—	.004 / .005		22000 / 27500		22000 / 27500
9	Small Vehicles, Fuel, Etc.	Fig.2.18.L	7200 MI	—	.50		3600		3600
10a	Cleanup - Progressive	Fig.2.18.N	LS	—	—	08838			8838
b	Final		LS	—	—	00000			00000
						+3200			+3200
11	Superintendent 21.65/HR		18 Mo	3752.00	—	67536			67536
12	Clerk 6.00/HR		18 Mo	1040.00	—	18720			18720
13	Fence w/10' Wide Gate	Fig.2.18.H	600 LF	—	1.96			1176	1176
14	Project Sign		LS	—	—			500	500
15	Temp. Closures For Weather		LS	—	—	1886			2686
16	Window & Fixture Cleaning		LS	—	—			002	002
17	Building Permits		5.5 M	—	500.00		2750		2750
18	Special Insurance		5.5 M	—	1250.00		6875		6875
19	Scaffolding (Included In Concrete Work)								
20	Watchman Weekends		18 Mo	1152.00	—	207 36			207 36
	SUB TOTAL					132019	123914	4676	260609
	FB 46% On Labor					128819			60722
	TOTAL					128819	116368		3097120

Figure 4.21a

PRICE OUT

JOB: **AIRCRAFT PARTS WAREHOUSE**
FLOOR AREA _____ COST PER SQ. FT. _____
LOCATION: **BOULDER, CA**
BID DATE: **JULY 7, 1982**
SHEET **2** OF **7**
DATE: **JUNE 15, 1982**

ITEM	DESCRIPTION	REF	QUANTITY	LABOR	SUB	MATERIAL	LABOR	SUB CONT.	TOTAL
	DEMOLITION (SITE CLEAR'G)								
1	Concrete Lined Ditch 4" Thick	Fig.2.20.G	1200 SF		.35			420	420
2	Pine Tree 16" Diam. & Stumps		20 EA		159.05			3182	3182
3	One Story Conc. Block Bld'g	See Work Sheet	800 SF		1.72			1376	1376
4	Slab & Footings	Fig.2.20.i	19 CY		98.31			1868	1868
5	Chain Link Fence 6' High	Fig.2.20.0	310 LF		1.30			403	403
6	Conc. Gutter Type Curb	Fig.2.20.h	280 LF		2.07			579	579
7	Saw Cut Conc. 1½" Deep	Fig.2.20.a	20 LF		2.50			50	50
8	Conc. Sidewalk 4" Thick	Fig.2.20.G	900 SF		.42			378	378
9	Sewer MH 8" Deep, 64 CF	Fig.2.20.0	1 EA		194.00			194	194
10	AC Paving 3" Thick	Fig.2.20.J	1500 SF		.16			240	240
								8690	8690
	Subcontractor Markup Say 20%							1738	1738
								10428	
	EARTHWORK								
1	Strip & Haul Off Site 4 Miles		1900 CY		2.67			5073	5073
2	Cut, Drift & Compact On Site		2844 CY		1.60			4550	4550
3	Scarify & Compact Overall		296,000 SF		.015			4440	4440
4	Import Fill Material		21,422 CY		4.37			93614	93614
5	Spread & Compact The Import		21,422 CY		.95			20350	20350
6	Fine Grade Overall		296,000 SF		.018			5328	5328
								133355	133355
	Subcontractor Markup Say 15%							20003	20003
									153358

Figure 4.21b

218

PRICE OUT

JOB: __AIRCRAFT PARTS WAREHOUSE__   COST PER SQ. FT. _____   FLOOR AREA _____   LOCATION: __BOULDER, CA.__   BID DATE __JULY 7, 1982__

CARPENTERS 19.65/HR
CEMENT MASONS 18.74/HR
LABORERS 16.37/HR

ITEM	DESCRIPTION		QUANTITY	LABOR	MATL	LABOR	MATERIAL	SUB CONT.	TOTAL
	__C.I.P. CONCRETE__								
⑬	1 LAYOUT, BATTERBOARDS, ETC.	Fig.2.21.A	LS	—	—	3062 / 3402	1100 / 1414		4162 / 4846
	2 MASS FOOTINGS								
⑭	a MACHINE EXC.		476 CY	—	3.33		1585		1585
	b HAND EXC.	Fig.2.21.C	9500 SF	.28	—	2660			2660
	c FORMWORK		6630 SF	2.28	.60	15116	3978		19094
⑮	d FORM DEPRESSION FOR P.C. PL.	*	2040 LF	.57	.25	1163	510		1673
	e DISPOSAL EXCESS DIRT		476 CY	—	2.82		1347		1347
⑯	f FORM CONST. JTS. IN FOOTINGS	*	8 EA	72.50	17.00	580	136		716
	g CONCRETE (POUR W/ SLAB)								
	h POINT & PATCH (200 SF/HR)	Fig.2.21.Z	7300 SF	.095	.05	694	365		1059
	i RUB & GRIND (60 SF/HR)	Fig.2.21.CC	6000 SF	.31	.09	1860	540		2400
	3 SPREAD FOOTINGS								
	a MACHINE EXC. (SAME AS 2a)		84 CY	—	3.33		280		280
	b HAND EXC. (SAME AS 2b)		950 SF	.28	3.22	266 / 85			266 / 136
	c FORMWORK – B 2 USES	Fig.2.21.H	380 LF	3.22	.66	885	251		1136
	d CONCRETE 3000 psi-1" Fig.2.21.5 & 2.21.T		31 CY	8.36	66.15	259	2051		2310
	e DISPOSAL (SAME AS 2E)		45 CY	—	2.83		127		127
	f BACKFILL CONDITION D	Fig.2.21.D	39 CY	8.18	3.15	319	146		465
	4 FOUNDATION STEM WALL								
	a FORMWORK 3' HIGH – 2 USES	Fig.2.21.J	2880 SF	2.12	1.73	6156	1664		7820
	b CONCRETE 3000 psi-1" Fig.2.21.5 & 2.21.T		31 CY	8.36	66.15	259	2051		2310
	c POINT & PATCH 180 SF/HR	Fig.2.21.Z	2280 SF	.10	.05	228	114		342
	5 12" THK. SLAB ON GRD. IN WAREHOUSE								
⑰	a MACHINE EXC. PILE CAPS	SEE 2a	930 CY	—	3.33	3097	3097		3097
	b DISPOSAL FROM PILES & CAPS	SEE 2e	1990 CY	—	2.83	5632	5632		5632
	c HAND EXC. AROUND PILES	4 PILES/HR	1767 EA	4.02		7227			7227
	d FINE GRADE	Fig.2.21.1	156,000 SF	.04	.023	6240	3588		9828
	e FORMWORK @ JOINTS 12"- 4 USES	Fig.2.21.AA	6480 LF	2.18	.42	14126	2722		16848
	f SET SCREEDS	Fig.2.21.5 & 2.21.T	156,000 SF	.112	.044	17472	5864		24336
⑱	g CONCRETE 3000 psi	Fig.2.21.5 & 2.21.T	8400 CY	6.19	20.23	57036	58932		64976
	__CARRY FORWARD__					1359948	6288424		7643372

Figure 4.21c

PRICE OUT

JOB __AIRCRAFT PARTS WAREHOUSE__   BID DATE _____   DATE __JULY 7, 1982__   SHEET __4__ OF __7__

FLOOR AREA _____   COST PER SQ. FT. _____   LOCATION _____   __JUNE 18, 1982__

ITEM	DESCRIPTION	REF	QUANTITY	LABOR	MAT'L	LABOR	MATERIAL	SUB CONT.	TOTAL
	C.I.P. CONCRETE		FORWARDED			1 359 48	628424		1 643 72
	h EXPAN. JOINTS & SEALER ¾" x 1½"	FIG.2.21.FF	1560 LF	.78	.87	1217	1357		2574
	i FINISH & CURE	FIG.2.21.BB	156,000 SF	.25 / .025	.025	39000 / 39000	3900		39700 / 42900
6	4" THK. SLAB ON GRD. IN OFFICE								
	a FINE GRADE 200 SF/HR	FIG.2.21.C	12,000 SF	.08	—	960			960
	b 1st 2" LAYER SAND - MAT'L	FIG.2.21.EE	90 CY	—	11⁶⁵		1049		1049
	LABOR & EQUIP.	FIG.2.21.DD	90 CY	11²⁴	11⁸⁰	1012	1062		2074
	c 6 MIL. POLY. MEMBRANE	FIG.2.21.EE	13,200 SF	.033	.033	436	436		872
	d 2nd 2" LAYER SAND - MAT'L	FIG.2.21.EG	90 CY	—	11⁶⁵		1049		1049
	LABOR & EQUIP.	FIG.2.21.DD	90 CY	11²⁴	11⁸⁰	1012	1062		2074
	e SET SCREEDS	FIG.2.21.AA	12,000 SF	.085	.033	1020	396		1416
	f CONCRETE 2500 PSI-1	FIG.2.21.S &2.21.T	166 CY	7⁶³-L	58⁰⁰	1267	9628		10895
	PUMPING	FIG.2.21.T	133 CY	—	10⁹⁴		1455		1455
	g FINISH & CURE 600 SF/DAY	FIG.2.21.BB	12,000 SF	.31	.025	3720	300		4020
7	STEPS ON GRD.	*	120 LF	8⁸⁰	3⁴⁰	1056	408		1464
8	SAND, FILL & TEMP. SLAB IN LOADING DOCK RAMPS	*	3 EA	262³³	145⁰⁰	787	435		1222
9	COLUMNS BETWEEN PRECAST PANELS								
	a FORMWORK 18' HIGH. 5 USES	FIG.2.21.K	6,120 SF	4⁰⁵ / 4⁴⁰	.75 / .96	24480 / 26920	4570 / 5508		27070 / 32436
	b CHAMFERS	*	8,160 LF	.25	.09	2040	734		2774
	c CONCRETE 3500 PSI-¾"	FIG.2.21.S &2.21.T	83 CY	8⁸⁶	74²⁵	735	6217		6952
	PUMPING 20CY/4 HR RATE	FIG.2.21.T	83 CY	—	40⁶⁵		3374		3374
	d POINT & PATCH 100SF/HR	FIG.2.21.Z	6,120 SF	.19	.09	1163	551		1714
	e RUB & GRIND/SACK 40SF/HR	FIG.2.21.CC	6,120 SF	.47	.14	2876	857		3733
10	PITS FOR ADJ. LOADING RAMPS W/ TEMP. FILL *	FIG.3.1.EA	3 EA	580⁰⁰	300⁰⁰	1740	900		2640
11	CURB GUTTER TYPE - STRAIGHT	FIG.3.1.EA	350 LF	—	9²⁴			3234	3234
12	CURB PLAIN TYPE - STRAIGHT	FIG.3.1.EA	580 LF	—	8⁴⁹			4924	4924
13	SIDEWALKS	FIG.3.1.DA	1550 SF	—	1⁹⁰			2945	2945
	ADD ROCK SALT	FIG.3.1.DA	950 SF	—	.25			238	238
14	CATCH BASINS 4'x4'x5'	FIG.2.21.Y	2 EA	617⁰⁰	308⁰⁰	1234	616		1850
	FB 4670	FIG.2.27.A				22 4151	66 9718	11 341	90 5210
									10 3109
									+100 8319
						21 8498	65 3924		97 2931

Figure 4.21d

PRICE-OUT

JOB	AIRCRAFT PARTS WAREHOUSE	BID DATE July 7, 1982
FLOOR AREA	COST PER SQ. FT.	LOCATION Boulder, CA.
		DATE June 21, 1982

PRECAST (TILT-UP) CONCRETE WORK

ITEM	DESCRIPTION	QUANTITY	LABOR	MATL	LABOR	MATERIAL	SUB CONT.	TOTAL
1	Engineering (dwgs. & calcs.)	105 pts		$18.00			1890	1890
2	Bond breaker (18 s.f./hr per man)	50,400 SF	.025	$.03	1260	1512		2772
(17) 3	Form edges (8 uses)	9240 LF	1.84	.21	17002	2310		19312
4	Chamfer strips (100 LF./hr.)	16,380 LF	.197	$.08	3227	1310		4537
5	Set bolts for roof ledgers (8/hr)	700 ea.	2.46	-	1722			1722
6	Set lift inserts (6/hr.)	630 ea.	3.28	$3.95	2066	2489		4555
7	Set brace inserts (7/hr.)	420 ea.	2.81	$3.33	1180	1399		2579
8	Set channel door frames (2 hrs. ea.)	12 ea.	39.30	-	472			472
(29) 9	Concrete 3500 p.s.i. 3/4" agg.	1280 CY	6.30	78.29	8064	100211		108275
10	Finish & cure (labor + 8.47%)	50,400 SF	.27	.025	13608	1260		14868
(21) 11	Erection - crane & riggers (Fig 2.21.bb)	105 pts	-	$347.00			36435	36435
12	Braces - install & remove (1½ hr. ea.)	210 ea.	29.48	$20.00	6191	4200		10391
13	Leveling, grouting under panels (1½ hr. ea.)	105 ea.	28.11	11.39	2952	1196		4148
14	Point & Patch (1 hr. ea.)	105 ea.	18.74	5.70	1968	599		2567
15	Scaffolding	105 hrs	1.00	1.50	105	158		263
	FB 46% of labor				59817	116644	38325	214782
								27516
								242302

SUMMARY (compare to sub bids)

			LABOR	MATERIAL	SUB CONT.	TOTAL
Casting - (items #1 thru #10)			48601	110491	1890	160982
FB 46% of labor				22356		22356
						183338
ERECTING - (item #11)					36435	36435
SECURING - (items #12 - #15)			11216	6153		17369
FB 46% of labor				5160		5160
						22529
TOTAL ESTIMATE PRECAST CONCRETE WORK						242302

Figure 4.21e

ITEM	DESCRIPTION		QUANTITY	LABOR	MATL	LABOR	MATERIAL	SUB CONT.	TOTAL
	INSTALL MISC. METAL								
1	Pipe guard posts incl. exc. & conc.		8 ea.	16.37	15.00	131	120		251
2	Angle door sills	1/2 mhr	12 ea.	9.83	—	118			118
3	Channel door frames (included in precast conc. work)								
4	Angle in edge of dock	25 LF/mhr	340 LF	.79	—	269			269
5	Pipe sleeves for handrail	1/4 mhr	6 ea.	4.91	—	29			29
6	Dock bumpers	1 mhr	30 ea.	19.65	—	590			590
7	Angle in adj. loading ramp pits	10LF/mhr	196 LF	1.97	—	386			386
8	Ladders to roof	2 mhr	3 ea.	39.30	—	118			118
9	Roof scuttles	6 mhr	4 ea.	117.90	—	472			472
10	Catch basin frames w covers	3/4 mhr	2 ea.	14.74	—	29			29
11	Safety nosing on steps	10 LF/mhr	24 L.F.	1.97	—	47			47
12	Anchor plates in top of cols.	2/3 mhr	72 ea.	13.17	—	948			948
13	Ledger angles	25 LF/mhr	380 L.F.	.79	—	300			300
14	Misc. hoisting equipment		20 hr.	—	65.00		1300		1300
						3437	1420		4857
	FB 46% on labor								1581
									6438
	CAULKING & SEALANTS								
1	Metal door frames - latex Fig. 2.26.c 1/2 × 1/2		80 LF	.52	.12	42	10		52
2	Tilt-up concrete control Jts. 3/4 × 3/4		240 LF	.59	.23	142	55		197
a	Backer rods - 3/4" Fig 2.26.b		240 LF	.39	.52	94	125		219
3	Masonry wall control Jts. 3/4 × 3/4		80 LF	.59	.23	47	18		65
4	Gypsum BD. - elastomeric 1/4 × 1/4		300 LF	.46	.03	138	9		147
a	Backer rods - 1/4" Fig 2.26.b		96 LF	.36	.18	35	17		52
5	Miscellaneous - Polysulphide 1/4 × 1/2		150 LF	1.12	.23	168	35		203
						666	269		935
	FB 46% on labor								306
									1241

Figure 4.21f

222

JOB AIRCRAFT PARTS WAREHOUSE **LOCATION** BOULDER, CA. **BID DATE** July 7, 1982 **SHEET** 7 **OF** 7

FLOOR AREA _____ **COST PER SQ. FT.** _____ **DATE** June 21, 1982

ITEM	DESCRIPTION		QUANTITY	LABOR	MATL	LABOR	MATERIAL	SUB CONT.	TOTAL
	CARPENTRY								
1	Metal door frames, grouted	2 hr.ea.	9 ea.	39.30	4.00	354	36	—	390
2	Metal doors 3'x7' s.c.	Fig 2.24.c	4 ea.	27.32	—	109		—	109
3	Wood doors 28"x7'	Fig 2.24.b	5 ea.	34.83	—	174		—	174
4a	Door hardware – thresholds	Fig 2.24.d	2 ea.	19.65	—	39		—	39
b	closers	"	4 ea.	58.95	—	236		—	236
c	exit devices	"	2 ea.	78.61	—	157		—	157
d	kick plates	"	2 ea.	14.74	—	29		—	29
e	weatherstrips	"	2 ea.	39.31	—	79		—	79
5a	Nailers on roof 3x4, 2x6	18 bf/hr.	180 LF.	1.09	.90	196	162		358
b	3x8, 2x12	16 bf/hr.	360 LF.	1.23	1.10	443	396		839
6	Base cabinet - std. grade	Fig 2.24.e	8 LF.	9.86	*10.00	79		560	639
7	overhead wall cabinet	"	20 SF.	2.11	*15.00	42		300	342
8	Misc. rough hardware		LS		LS		75		75
						1937	669	860	3466
	FB 46% on labor								891
									4357
	MASONRY								
1	Concrete block 8" sol., grout	Fig. 3.11a	3600 SF		6.16				
a	add for fluted				1.10				
b	add for color				.20				
	TOTAL		3600 SF		7.46			26856	
c	add for shoring in windows/doors		1 ea.		*$1200			840	
	SUB TOTAL							27696	
d	add mileage for out-of-town sub				*10%			2770	30466
								30466	96
	NETWORK ANALYSIS								
		Fig 2.18.P 5,500.00 $			.0008 /ft	4400	4400		4400
						5500	5500		5500

Figure 4.21g

Specifications Summary and Bid Sheet

Project: AIRCRAFT PARTS WAREHOUSE   **Bid Date:** JULY 7, 1982

		Base	Alt. A	Alt. B	Alt. C
01100	General Conditions	309,120	—	—	—
01311	Network Analysis	4,400	—	—	—
02110	Demolition	—	—	—	—
02200	Earthwork	155,000	—	—	502
02300	Prestressed conc piling	883,500	—	—	—
02444	Chain link fencing	28,160	—	—	—
02800	Landscaping & sprinklers	19,450	—	—	—
03300	C.I.P. concrete	972,931	—	(1,584)	37,240
03400	Precast concrete	242,302	—	—	—
03600	Reinforcing steel	194,240	—	—	—
02800	Asphalt concrete paving	67,200	—	—	(8,400)
04230	Concrete block masonry	30,466	—	—	—
05120	Structural steel	375,000	—	—	—
05320	Steel roof decking	151,151	—	—	—
05500	Miscellaneous metal	24,330	—	—	—
	Install miscellaneous metal	6,438	—	—	—
06000	Carpentry	4,357	—	—	—
07241	Roof insulation	96,700	—	—	—
07510	Roofing	77,000	—	—	—
07600	Sheet metal work	33,330	—	—	—
07951	Caulking & sealants	1,241	—	—	—
08110	Hol. metal doors & frames	2,800	—	—	—
08310	Sliding fire doors	36,575	—	—	—
08330	Steel studs & gyp board	24,500	—	—	—
08710	Finish hardware	2,000	—	—	—
09910	Painting	75,000	—	—	—
11675	Pallet storage racks	—	220,000	—	—
11871	Adjustable loading ramps	—	—	58,000	—
15000	Mechanical	795,550	—	—	—
16000	Electrical	387,450	—	4,200	—
	Subtotal	5,000,191	220,000	60,616	29,342
	Markup 9%	450,017	19,800	5,455	2,641
	Contingency (Cut)	(27,000)	—	—	—
	Subtotal	5,423,208	239,800	66,071	31,983
	Bond Cost	27,116	1,199	330	160
	Bid Amount	5,450,324	240,999	66,401	32,143

Figure 4.21h

224

Index

Insurance, special risk, 164
Item
    definition, 9, 44

J

Job
    definition, 44
Judgment, 2
    calculated risk, 10-11
    in cost estimating, 9-10, 29
    identifying uncertain costs,
      10
    mistakes of, 29
    in rating desirability, 39-40,
      214

L

Labor costs
    carpentry, 50, 130-133, 135,
      138, 139, 140
    concrete work, 95, 112, 114
    crew-hours and pay scales, 53
    demolition, 88, 92
    effect of inflation on, 34
    equipment operators, 36, 56
    figuring, 54
    gross pay, 56
    lump sum, 56
    miscellaneous metal, 142
    mobilization time, 32-33
    price-out, 53-54, 55-56
    project wages, 35
    projected increases, 33, 34-35,
      186-187
    simple production/man-hour
      equation, 56
    simple unit cost, 56
    standby time, 33
    time lag allowance, 33
    See also Fringe benefits
Layout structures, 78, 83, 211
Left-on-the-table
    definition, 44
Let
    definition, 44
Lift inserts, 127
Liquidated damages, 85
Lump sum, 44

M

Main division, 44

Manholes
    price examples, 120
Markup
    adjusting for project de-
      sirability, 172, 216
    definition, 44
    finder scales, 172, 214
    and size of project, 170
Markup Finder Scale, 172
Masonry, 206
    budgeting, 160-161
    in-progress schedule, 209
    sample price-out, 208, 223
    unit prices, 161
Material costs
    allowances, 56
    average vs. incremental
      estimating, 36
    extensions, 57
Material suppliers
    rules for dealing with,
      162-163
Materials
    concrete formwork, 48,
      112-113
    earthen, 48
    effect of inflation on, 34, 36
    estimating techniques, 52,
      57, 58
    price fluctuations, 57-58
Mathematics
    decimal fraction of a foot, 50
    definition, 44
    law of proportions, 51
    working with decimals,
      52-53
Mechanical work, 209
Metal work, miscellaneous, 209
    installation, 206
    sample price-out, 142, 207,
      222
    subcontractor responsibil-
      ities, 142
Millwork, 32, 129, 139, 140
Mistakes in estimating
    in arithmetic, 30
    of judgment, 29
    of omission, 29-30
    overestimating, 39
    preventive measures, 30-31,
      53
    slips, 30